# TIN AND ITS ALLOYS AND COMPOUNDS

# ELLIS HORWOOD SERIES IN INDUSTRIAL METALS

*Series Editor:* E. G. WEST, OBE, Metallurgical Consultant, and former Director of the Copper Development Association, London

A systematic course of study on metallurgy and materials science for lecturers and students, to graduate level, as well as engineers in production and design, management scientists, researchers and other personnel in many areas of industry.

**TIN AND ITS ALLOYS AND COMPOUNDS**
B. T. K. BARRY and C. J. THWAITES, International Tin Research Institute, Middlesex
**COBALT AND ITS ALLOYS**
W. BETTERIDGE, former Chief Scientist, International Nickel Limited, UK
**PRINCIPLES OF HYDROMETALLURGICAL EXTRACTION AND RECLAIMATION**
E. JACKSON, Principal Lecturer in Metallurgy, Sheffield City Polytechnic, Sheffield
**COPPER AND ITS ALLOYS**
E. G. WEST, OBE, former Director, Copper Development Association, London
**BASIC CORROSION AND OXIDATION**
J. M. WEST, Department of Metallurgy, University of Sheffield

# TIN AND ITS ALLOYS
# AND COMPOUNDS

B. T. K. BARRY, B.Sc.(Eng.)., A.R.S.M., Ph.D., D.I.C., F.I.M., M.I.M.M., C.Eng.
Assistant Director, International Tin Research Institute

and

C. J. THWAITES, A.R.S.M. M.Sc., D.Sc., F.I.M., F.I.C.T., F.I.M.F., M.B.I.F.
Research Manager, International Tin Research Institute

ELLIS HORWOOD LIMITED
Publishers · Chichester

Halsted Press: a division of
JOHN WILEY & SONS
New York · Brisbane · Chichester · Ontario

First published in 1983 by
**ELLIS HORWOOD LIMITED**
Market Cross House, Cooper Street, Chichester, West Sussex, PO19 1EB, England

*The publisher's colophon is reproduced from James Gillison's drawing of the ancient Market Cross, Chichester.*

**Distributors:**

*Australia, New Zealand, South-east Asia:*
Jacaranda-Wiley Ltd., Jacaranda Press,
JOHN WILEY & SONS INC.,
G.P.O. Box 859, Brisbane, Queensland 40001, Australia

*Canada:*
JOHN WILEY & SONS CANADA LIMITED
22 Worcester Road, Rexdale, Ontario, Canada.

*Europe, Africa:*
JOHN WILEY & SONS LIMITED
Baffins Lane, Chichester, West Sussex, England.

*North and South America and the rest of the world:*
Halsted Press: a division of
JOHN WILEY & SONS
605 Third Avenue, New York, N.Y. 10016, U.S.A.

©1983 B. T. K. Barry and C. J. Thwaites/Ellis Horwood Limited

**British Library Cataloguing in Publication Data**
Barry, B. T. K.
Tin and its alloys and compounds. − (Ellis Horwood series in industrial metals)
1. Tin
I. Title   II. Thwaites, C. J.
669'.6   TN793
**Library of Congress Card No.** 83–12760

ISBN 0–85312–649–6 (Ellis Horwood Limited)
ISBN 0–470–27480–8 (Halsted Press)

Typeset in Press Roman by Ellis Horwood Limited.
Printed in Great Britain by R. J. Acford, Chichester.

# Table of Contents

## Chapter 4 – Tin and Tin Alloy Applications

## Chapter 5 – Solder Alloys

## Chapter 6 – Bearing Alloys

## Chapter 7 – Melting, Casting and Fabrication of Tin and Tin Alloys

## Chapter 8 – Tin and Tin Alloy Coatings

# Authors' Preface

Tin is a technologist's metal, in the sense that whilst its manifold applications reach into all aspects of life, nevertheless, most people are usually unaware of its presence.

Long before history came to be written, tin was associated with the technological advance of mankind and archaeological evidence confirms that tin was one of the earliest metals known to human society. The main reason for the importance of tin in early times was the discovery of its hardening effect on copper to form bronze. This alloy could be sharpened to give a cutting edge and hence could be fabricated into efficient tools and weapons. From this early Bronze Age to the Space Age of today, tin has continued to play an important role out of all proportion to the gross tonnage produced and used.

The main uses of tin are based on empirical knowledge accumulated over many years. Nevertheless there is a definite relationship between the established uses of tin and its physical and chemical properties, however little these aspects were realised at the time when tin was first adopted for a given use. In the latest applications of tin, new or improved industrial uses have followed from the determination of its properties and their subsequent exploitation.

Important properties of tin and tin-rich alloys which are commercially exploited include their low melting point and their ability, when molten, to wet other metals, their excellent castability, and their outstanding anti-frictional behaviour in the presence of oil. These characteristics, individually or combined, are of especial importance in applications such as coatings, solders, cast alloys and bearings.

In most of its applications, tin is employed in conjunction with other metals or materials in ways where its special properties can effect some significant improvement to the production or performance of the system: indeed, there are few uses where tin is the major constituent. The most important example of such use is pewter which normally contains over ninety per cent of tin and may be seen in many households, either as decorative articles or as utensils such as tankards and goblets. Even tinplate, which represents the major tonnage consumed, contains on average, less than half of one per cent of tin by weight, whilst solder, which consumes nearly thirty per cent, rarely contains more than sixty per cent tin by weight.

These are well established uses for tin, as are bearing metals, plated coatings, fusible alloys and bronzes; but there are a host of other minor metallurgical uses, including newer developments such as the use of tin in cast iron, in sintered iron parts, and in alloy powder compacts, as well as many lesser known applications covered by the text in this book.

Tin is also very important as an industrial chemical, and a characteristic of special significance is its capacity to form organotin compounds which, in their turn are the basis of many industrially useful chemicals.

The trading pattern of tin is also interesting in that almost ninety-five per cent of the world's tin output is used in the highly industrialised countries, almost all of which have no tin deposits of economic importance; conversely, the major tin ore producers are developing countries and as such they have no significant tin-using industries. Moreover, for most of the larger producers, tin exports represent a substantial source of revenue and make a major contribution to the well-being of their population. For these reasons, tin is a very important item of international trade.

It can be seen, therefore, that tin in all its applications and in the economics of its use, presents a very different picture from that of most other metals. It is believed that this book will provide a balanced picture of the technology of tin and its uses, with some consideration of the economic aspects involved. A guide to further reading for more detailed studies has been included with each chapter.

The authors are grateful to the Director of the International Tin Research Institute and the International Tin Research Council for permission to prepare this book. Photographic illustrations not otherwise credited are the copyright of the International Tin Research Institute.

B.T.K.B., C.J.T.

# 1

# Physical and Chemical Properties

Tin is used in a wide variety and range of industrial products and processes, as described and illustrated in this monograph. Nevertheless, despite its multiplicity of uses, the total world consumption of tin is relatively small, being only of the order of 200 000 tonnes per annum. This contrasts with annual consumption figures for the major base metals like copper, lead and zinc of between 3 million and 9 million tonnes annually.

Although it is not found in its native state, tin is one of the oldest metals known to have been produced and worked by man. Bronze, an alloy of copper and tin, dates from prehistoric times, and its importance as a medium for making weapons and other useful articles has led archaeologists to use the term Bronze Age to define the period when this metal reigned supreme. Yet despite its antiquity, tin remains today a vital component of advanced technology as well as an important item of commerce.

Tin owes its place in industry to certain basic properties. As a metal its most important characteristics are its low melting point, the ability to form alloys with most other metals, non-toxicity and resistance to corrosion, allied with good appearance. Tin and its alloys also have an excellent capability for retaining an oil film, hence their value in bearings. In chemicals, the value of tin lies in the wide range of compounds, of very different characteristics, that can be synthesised.

In its applications as a metal, tin is almost always used in partnership with other metals, either as a coating or as an alloying element. This is because its intrinsic softness prevents it from being used as a structural material unless strengthened by the addition of alloying elements.

Tin chemicals have many and varied applications as shown in Chapter 10. Both inorganic and organometallic tin compounds can be formed and find industrial use in electroplating, in pharmaceutical and agricultural products, in plastics and ceramics, and in many other industrial chemical fields. The formation of stable tin-to-carbon bonds in molecular structures has provided the basis for the large and growing field of organotin chemistry.

## 1.1 PHYSICAL PROPERTIES

Tin is a relatively soft and ductile metal with a 'white' colour similar to that of silver. A familiar characteristic of tin is the cracking noise emitted during deformation, called 'cry' of tin. This is especially noticeable when coarse-grained cast sticks are bent and is due to the plastic deformation causing twinning of the crystals on the 301 planes.

The properties of tin listed in this chapter represent a selection of data chiefly of metallurgical interest. The data are based on the most recent values, but fuller details can be found in standard reference works.

### Atomic and Nuclear Properties

The chemical symbol for tin is Sn, derived from the Latin name *stannum*. It is element number 50 in the periodic classification of the elements. The atomic properties are shown in Table 1.

### Table 1

Atomic properties

| | |
|---|---|
| Atomic number | 50 |
| Atomic mass (weight) | 118.69 |
| Atomic radius (12-fold coordination) | $1.58 \times 10^{-1}$ nm |
| Ionic radius $(Sn^{2+})$ | $0.93 \times 10^{-1}$ nm |
| $\quad\quad\quad (Sn^{4+})$ | $0.71 \times 10^{-1}$ nm |
| Valencies | 2 and 4 |
| Density $(\beta$ tin) | |
| $\quad$ (measured at 15 °C) | 7.29 g/cm³ |
| $\quad$ (by X-ray at 25 °C) | 7.281 g/cm³ |
| $\quad (\alpha$ tin) | |
| $\quad$ (measured at 15 °C) | 5.77 g/cm³ |
| $\quad$ (by X-ray at room temperature) | 5.765 g/cm³ |
| $\quad$ (liquid) | |
| $\quad$ (measured at melting point) | 6.976 g/cm³ |

Although the atomic mass of tin is quoted as 118.69, this is the mean derived from ten naturally occurring isotopes whose abundance varies from below 1 per cent to 32 per cent for that nearest to the mean value. There are also 20 radioactive isotopes that have been produced up to the present time, whose half-lives extend up to 76 years. These data are collated in Table 2.

## Table 2

Isotopes of tin

| (a) Stable isotopes | | (b) Unstable isotopes | |
| --- | --- | --- | --- |
| Atomic weights | % abundance | Atomic weights | Half life |
| 112 | 0.96 | 108 | 9 min. |
| 114 | 0.66 | 109 | 18.1 min. |
| 115 | 0.35 | 110 | 4 h |
| 116 | 14.30 | 111 | 35 min. |
| 117 | 7.61 | 113m | 20 min. |
| 118 | 24.03 | 113 | 115 d |
| 119 | 8.58 | 117m | 14 d |
| 120 | 32.85 | 119m | 250 d |
| 122 | 4.72 | 121m | 76 y |
| 124 | 5.94 | 121 | 27 h |
| | | 123 | 125 d |
| | | 123m | 42 min. |
| | | 125m | 8 min. |
| | | 125 | 9 d |
| | | 126 | $10^5$ y |
| | | 127 | 2 h |
| | | 128 | 59 min. |
| | | 130 | 2.6 min. |
| | | 131 | 3.4 min. |
| | | 132 | 2.2 min. |

### Crystallography and X-ray data

Tin in its usual form (white or $\beta$ tin) has a body-centred tetragonal crystal structure. On the other hand, a low temperature form (grey or $\alpha$ tin) exists which is a high resistivity, semiconducting material having a diamond-type cubic structure. This allotrope theoretically is stable below 13 °C but in practice is only readily formed at around $-40$ °C and the transformation is severely retarded by the presence of alloying elements or even trace impurities.

The density of grey ($\alpha$) tin is much lower than that of $\beta$ tin (see Table 1) so there is a disruptive volume change during transformation (tin 'pest'). Grey tin and the transformation to $\alpha$ from normal $\beta$ tin has not found a commercial application and may be regarded as a rather rare metallurgical curiosity. The characteristic atomic structure values are given in Table 3.

For the research metallurgist, a commonly used identification technique is that of taking X-ray diffraction patterns and therefore the salient X-ray data for tin (also given in Table 3) are of importance.

Table 3

Crystallography and X-ray data

| | |
|---|---|
| Structure | |
| ($\beta$ tin) | body-centred tetragonal |
| ($\alpha$ tin) | face-centred cubic |
| Lattice type | |
| ($\beta$ tin) | A5 |
| ($\alpha$ tin) | A4 |
| Lattice constants | |
| ($\beta$ tin) | $a = 58.315 \times 10^{-2}$ nm |
| | $c = 31.815 \times 10^{-2}$ nm |
| | $c/a = 0.5456$ |
| ($\alpha$ tin) | $a = 64.912 \times 10^{-2}$ nm |
| Atoms per unit cell | |
| ($\beta$ tin) | 4 |
| ($\alpha$ tin) | 8 |
| Transformation temperature ($\alpha \leftrightarrows \beta$) | 13.2 °C |
| Twinning plane ($\beta$ tin) | (301) |
| Emission spectrum strongest lines | $4.9502 \times 10^{-2}$ nm ($K\alpha_2$) |
| | $4.9056 \times 10^{-2}$ nm ($K\alpha_1$) |
| | $35.995 \times 10^{-2}$ nm ($L\alpha_1$) |
| | $33.848 \times 10^{-2}$ nm ($L\beta_1$) |
| | $31.743 \times 10^{-2}$ nm ($L\beta_2$) |
| Excitation potentials | 29.1 kV (K) |
| | 4.49 kV (L) |
| | 0.88 kV (M) |
| | 0.13 kV (N) |
| Diffraction lattice (d) spacings | |
| for $\beta$ tin | $29.15 \times 10^{-2}$ nm (100) |
| (intensities as ratio of strongest | $27.93 \times 10^{-2}$ nm (90) |
| line are in brackets) | $20.62 \times 10^{-2}$ nm (34) |
| | $20.17 \times 10^{-2}$ nm (74) |
| | $16.59 \times 10^{-2}$ nm (17) |
| | $14.84 \times 10^{-2}$ nm (23) |
| | $14.58 \times 10^{-2}$ nm (13) |
| | $14.42 \times 10^{-2}$ nm (20) |
| | $13.04 \times 10^{-2}$ nm (15) |
| | $12.92 \times 10^{-2}$ nm (15) |
| | $12.05 \times 10^{-2}$ nm (20) |
| | $1.095 \times 10^{-2}$ nm (13) |
| for $\alpha$ tin | $37.3 \times 10^{-2}$ nm (100) |
| | $22.8 \times 10^{-2}$ nm (80) |
| | $19.5 \times 10^{-2}$ nm (70) |
| | $16.1 \times 10^{-2}$ nm (30) |
| | $14.8 \times 10^{-2}$ nm (50) |
| | $13.2 \times 10^{-2}$ nm (60) |
| | $12.4 \times 10^{-2}$ nm (40) |
| | $11.4 \times 10^{-2}$ nm (20) |
| | $10.9 \times 10^{-2}$ nm (40) |

**Thermal properties**

The low melting point of tin, at just above 232 °C, is made use of in its large-scale applications as a low-melting and joining metal and as a metallic coating. On the other hand, despite its low melting temperature, tin has a quite low vapour pressure even at temperatures in excess of 1000 °C, with a boiling point of about 2600 °C (Table 4). This property is exploited in some applications (see Chapter 4) but is a disadvantage in that steel melts contaminated with tin cannot be purified by simply raising the temperature: even vacuum melting at high temperatures is not a viable method of removing tin from steel.

**Table 4**

Thermal properties

| | |
|---|---|
| Fusion point | 231.88 °C |
| Boiling point | 2625 °C |
| Vapour pressure | |
|     at 727 °C | $7.4 \times 10^{-6}$ mm Hg |
|     at 1127 °C | $4.4 \times 10^{-2}$ mm Hg |
|     at 1527 °C | 5.60 mm Hg |
| Latent heat of fusion | 7.08 kJ/g atom |
| Latent heat of vaporisation | 296.4 kJ/g atom |
| Specific heat at 20 °C | 222 J/kg K |
| Normal entropy at 25 °C | 57.5 J/kmol |
| Thermal conductivity at 20 °C | 65 W/m K |
| Coefficient of expansion | |
|     (linear at 0 °C) | $19.9 \times 10^{-6}$ |
|     (linear at 100 °C) | $23.8 \times 10^{-6}$ |
|     (cubical at 0 °C) | $59.8 \times 10^{-6}$ |
|     (cubical from melting point to 400 °C) | $106 \times 10^{-6}$ |
|     (single crystal at 0 °C) | $28.4 \times 10^{-6}$ |
| | (parallel to c-axis) |
| | $15.8 \times 10^{-6}$ |
| | (normal to c-axis) |
| Surface tension at melting point | 544 mN/m |
| Viscosity at melting point | 1.85 mNs/m$^2$ |
| Expansion on melting | 2.3% |
| Gas solubility in liquid tin | |
|     (oxygen at 536 °C) | 0.00018% |
|     (oxygen at 750 °C) | 0.0049% |
|     (hydrogen at 1000 °C) | 0.04% |
|     (hydrogen at 1300 °C) | 0.36% |
|     (nitrogen) | very low |

Tin and tin-rich alloys exhibit extreme fluidity at just above the melting point as evidenced by the viscosity values shown. It will be noted that tin exhibits anisotropy of the thermal expansion properties in different crystallographic directions, and this may lead to steps appearing in a free surface at the boundaries between large grains in samples subjected to severe thermal cycles, as in whitemetal bearings.

A notable feature of tin is the very low solubility of gases even in liquid tin at up to quite high temperatures, so that gas pick-up during the melting of tin alloys does not occur (Table 4). On the other hand, the slight solubility of nitrogen and the reaction to form a nitride at very high temperature is made use of in a prototype nuclear reactor (see Chapter 11).

**Electrical properties**

Tin is a relatively poor electrical conductor compared with copper, namely 13 per cent IACS, and resistivity values are given in Table 5. The element becomes a superconductor at temperatures near absolute zero and, indirectly, use is made of this in the most commonly used commercial superconducting material, $Nb_3Sn$ (see Chapter 11). Other electrical properties are included in the table.

**Table 5**

Electrical properties

| | |
|---|---|
| Resistivity ($\beta$ tin) | |
| at 20 °C | 12.6 $\mu\Omega$cm |
| at 200 °C | 23.0 $\mu\Omega$cm |
| at 300 °C | 46.8 $\mu\Omega$cm |
| Resistivity ($\alpha$ tin) | |
| at 0 °C | 300 $\mu\Omega$cm |
| Superconducting transition temperature | 3.73 K |
| Photoelectric work function | 4.64 eV |
| Electrochemical equivalent | |
| ($Sn^{2+}$) | 0.61503 mg/C |
| ($Sn^{4+}$) | 0.30751 mg/C |
| Single electrode potential | $-0.52$ V |
| (against saturated calomel in dilute chloride solution) | |

## 1.2 MECHANICAL PROPERTIES

The low melting point of tin means that at normal ambient temperatures the metal is at nearly 60 per cent of its melting point on the absolute temperature scale, resulting in rather low mechanical strength at room temperature (Table 6), as would be expected from any metal tested at a temperature corresponding to such a high proportion of its absolute melting point, 505 K. Nevertheless the mechanical properties of tin are strongly dependent on impurity levels and the rate of testing, while the change in the available slip planes with decreasing temperature causes a sudden drop in ductility at about $-120\,°C$.

### Table 6

Mechanical properties

| | |
|---|---|
| Tensile strength (at 0.4 mm/mm/min) | |
| at 20 °C | 14.5 N/mm² |
| at 100 °C | 11.0 N/mm² |
| at 200 °C | 4.5 N/mm² |
| Shear strength at room temperature | 12.3 N/mm² |
| Young's modulus at 20 °C | 49.9 kN/mm² |
| Rigidity modulus at 20 °C | 18.4 kN/mm² |
| Bulk modulus at 20 °C | 58.2 kN/mm² |
| Poisson's ratio | 0.357 |
| Creep strength at 15 °C | |
| (approx. life at 2.3 N/mm²) | 170 days |
| (approx. life at 1.4 N/mm²) | 550 days |
| Fatigue strength for $10^8$ reversals at 15 °C | ±2.5 N/mm² |
| Hardness | |
| at 20 °C | 3.9 HB |
| at 100 °C | 2.3 HB |
| at 200 °C | 0.9 HB |

*Note:* tensile strength, hardness, etc., are very dependent on rate of loading

Because of the proximity of the melting point, pure tin recrystallises readily at room temperature. Consequently, unlike the majority of industrially used metals, only slight work-hardening occurs initially, followed by work-softening with further deformation, due to grain growth. This characteristic also applies to the tin-rich alloys such as tin–lead (soft solders – see Chapter 5) and pewter (tin–antimony–copper – see Chapter 4).

### 1.3 CHEMICAL PROPERTIES

The preparation and properties of the chemical compounds of tin are described in Chapter 10. This present chapter contains a selection of data relating to the chemical reactivity of metallic tin in contact with a limited range of gases and liquids. As with all such data, much depends on the ambient conditions such as temperature, presence or absence of moisture, strength of solutions, etc., so that the data quoted herein should be utilised for general guidance only. For detailed performance data a specialised reference or a competent authority should be consulted.

In general terms, pure tin is relatively resistant to reaction with, or attack by, gases and weak acids or alkalis at ambient temperatures. Thus, pure tin retains its silvery-white appearance remarkably well in both normal atmospheres and in quite aggressive industrially polluted atmospheres.

Tin occurs in Group IVB of the Periodic Table: the non-metal carbon and semi-metals silicon and germanium lie to one side of tin and the typically metallic element lead on the other side. Hence tin might be expected to show tendencies for some non-metallic behaviour. In its chemical reactions, tin is capable of exhibiting valencies of 2 or 4 (see Table 1).

Tin forms both divalent Sn(II) and tetravalent Sn(IV) salts. It enters into compounds as the anion or cation and most tin salts are freely soluble in water. It will also form complexes and a wide range of organometallic compounds with tin—carbon bonding; these compounds in general show various levels of biocidal activity (Chapter 10).

The most important chemical reactions of pure tin are given below, firstly with gases and secondly with liquid media.

## Gases

In dry oxygen and air at ambient temperatures, the rate of growth of the oxide film is slow (Fig. 1); it may be many years before coloration of the tin surface (due to a first order interference film) becomes noticeable, but the presence of moisture accelerates the oxidation. At temperatures of around 200 °C interference colours will form in an hour or so. Sulphur pollution of the atmosphere has to be at levels intolerable to human beings before the rate of tarnish formation on tin is significant. Reaction with water vapour to form tin oxide films occurs at 700 to 800 °C and, similarly, high temperatures cause reaction with carbon dioxide.

The halogen gases readily form stannic halides with tin, although the reaction with fluorine is slow below 100 °C.

Ammonia gas, hydrogen and nitrogen will not react directly with metallic tin; similarly tin is inert to organic acid vapours such as acetic acid unless they are hot.

## Water

Pure water does not react with tin so that tin is often used for water stills, sometimes as thick coatings on copper condensing coils. So-called 'block' tin stills are used for producing very high purity distilled water. Sea water attacks tin only very slowly; generally it produces pitting, together with a surface film of mixed stannous oxides and chlorides or stannic oxide with water of hydration.

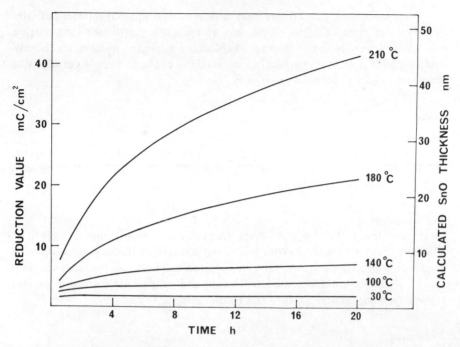

Fig. 1 — Rate of oxidation of a pure tin surface at various temperatures in a laboratory atmosphere. Oxide film thickness was determined by coulometric reduction.

### Dilute aqueous solutions

The behaviour of tin in dilute solutions depends on the single electrode potential, the hydrogen overpotential and the solubility of the reaction product. The presence or absence of oxygen or an oxidising agent, however, has a profound effect on the reactions of tin, as will be seen for certain acids in Table 7.

### Table 7

Corrosion of tin by strong acids in presence and absence of oxygen

| Acid | Weight loss (g/m² per day) | |
|---|---|---|
| | under hydrogen | under oxygen |
| 30% $HNO_3$ | 63 | 64 |
| 6% HCl | 6 | 1110 |
| 6% $H_2SO_4$ | 3.5 | 430 |
| 6% acetic acid | 1.5 | 230 |

Tin is resistant to general corrosion in near-neutral aqueous solutions of salts owing to the presence of an oxide film on the surface, although some pitting may occur especially in the presence of chlorides, sulphates, nitrates, etc. Highly oxidising salt solutions that are far from neutral (for example, sodium persulphate) attack tin rapidly to form an oxide film.

## Acids

### Halogen acids
Dilute hydrochloric acid attacks tin slowly in the absence of an oxidising agent, up to a strength of 6 M, to form halides (see Chapter 10). The other halogen acids (except hydrofluoric) attack tin readily when hot.

### Sulphuric acid
Dilute sulphuric acid (20 w/v per cent) reacts slowly at room temperature, but at 100 °C hydrogen is evolved. With increasing strength of this acid some reaction occurs at room temperature, but concentrated sulphuric acid attacks tin only slowly. Hydrogen sulphide, sulphur dioxide and sulphur are formed in various ratios depending on acid strength and temperature.

### Nitric acid
Dilute nitric acid reacts with tin to form soluble salts in dilute solutions, while the concentrated acid causes the formation of a surface film of hydrous stannic oxide (commonly called metastannic acid).

### Phosphoric acid
Phosphoric acid is far less reactive towards tin, probably because a film of protective phosphate is formed; similarly chromic acid passivates tin.

### Organic acids
Acetic acid is also much less reactive, especially in the absence of oxygen. Other organic acids such as oleic and stearic do not react at room temperature, hence the metal may be used to contain products based on these materials. When oxygen is absent, there is no reaction of tin with other weak acids such as citric and oxalic.

## Alkalis
Tin reacts with dilute solutions of sodium and potassium hydroxides and carbonates, especially when warm and in the presence of small amounts of oxidising agent, to form stannates or stannites. Saturated ammonia does not attack tin but dilute solutions react more like the other alkaline solutions of similar pH. Some amines may attack tin.

It is important to realise that the above statements refer to tin metal alone in contact with solutions. If the tin is only one of a number of metals in contact with each other and the solution, then galvanic effects may be expected and attack on the tin may be promoted. In particular, the presence of a metal on which the evolution of hydrogen can occur easily will allow the corrosion of tin to take place in both acids and alkalis in the absence of any oxidising agent. For example, tinned steel in which some of the steel is exposed by coating damage can be immersed in sodium hydroxide in the absence of air with a resulting dissolution of tin and evolution of hydrogen.

**Other liquids**
The majority of pure organic liquids are without action on tin. Examples are oils, refrigerants, petroleum, phenol, alcohols, esters, benzene, acetone and chlorinated hydrocarbon solvents. However, traces of decomposition product such as free acid, or of water, may result in some reaction occurring.

**FURTHER READING**

1.  *Properties of Tin,* International Tin Research Institute Publication No. 218.
2.  Smithells, C. J., *Metals Reference Book,* 5th edn., Butterworth, London, 1976.
3.  *Metals Handbook,* 8th edn., Vol. 2, American Society for Metals, Ohio, 1979.
4.  *Zinn-Taschenbuch,* Metall-Verlag, Berlin, 1975.
5.  Gmelin, L., *Handbuch der Anorganischen Chemie – Zinn (element),* Teil B, Verlag Chemie, Weinheim, 1981.
6.  *Tin versus Corrosion,* International Tin Research Institute Publication No. 510.
7.  Leidheiser, H., *The Corrosion of Copper, Tin and their Alloys,* Wiley, New York, 1971.

# 2

# Occurrence and Production

## 2.1 TIN MINERALS

Tin does not occur naturally as a metal. By far the most economically important tin mineral is cassiterite, a naturally occurring oxide of tin with the chemical formula $SnO_2$. In its purest form it contains 78.6 per cent tin. Cassiterite crystallises in the tetragonal system; it is a transparent mineral, usually brown or black in colour, with a hardness of 6 to 7 on the Mohs' scale. The specific gravity of cassiterite is around 7; this is high as compared with most rock-forming minerals with which it is associated and advantage is taken of this differential in the gravity separation processes used in the production of concentrates from the run-of-mine ore.

Although cassiterite is the only tin mineral of major commercial importance as a source of tin, there are a number of sulphidic tin minerals which are found associated with the ore bodies, especially in Bolivia where the tin ores are very complex. The sulphidic tin minerals include stannite, a complex sulphide of tin, copper and iron, sometimes known as tin pyrites. Its specific gravity is around 4.5, lower than cassiterite, and the tin content is about 22 to 27 per cent. Cylindrite is another complex sulphide mineral, containing also lead and antimony sulphides. Its specific gravity is about 5.4, intermediate between stannite and cassiterite.

Tin minerals are almost invariably found in association with granitic rocks, although the converse is not true, for only a fraction of the earth's granite rocks are stanniferous. Primary tin minerals occur chiefly in veins or lodes in the granitic country rock, although they may occasionally be found disseminated throughout the host rock itself. Tin-bearing lodes can have horizontal or vertical dimensions of several hundred metres, but are usually less than one metre in lateral width. Lode deposits may contain around one per cent of tin.

About 80 per cent of the world's tin is derived not from primary lodes but from unconsolidated secondary, or placer, deposits. Since cassiterite is very resistant to chemical and mechanical weathering, as well as having a high specific gravity, erosion of the host rock and the tin-bearing veins results in the concentration of

Fig. 2 – Gravel pump mining in which a high-pressure water jet removes alluvial deposits containing tin minerals which are carried by the water stream to the ore-treatment plant.

tin minerals at regions distant from the primary lodes. Many unconsolidated deposits consist of gravels, sands, silts and clay washed down by rivers and streams in times past and deposited in valleys or under the sea. Other such deposits are found on hillsides and are composed of rock debris, which has not been transported by water to any great extent. Placer deposits may contain as little as 0.015 per cent of tin.

## 2.2 MINING METHODS

The cheapest and most economical ways of mining these unconsolidated deposits are by dredging, by hydraulic methods or by open-cut mining. The workable tin deposits in the Far Eastern tin field are mostly of this type, although there is one underground mine in Malaysia. Similar techniques are used in the African tin mines, whilst in Australia both alluvial and lode deposits are mined. In Bolivia, the world's fourth largest producer, the bulk of production is by underground mining of primary lode deposits. Moreover, the ore is very complex which makes extraction of the metal more expensive. For these reasons, Bolivia is a high-cost producer as compared with the other major producing countries.

### Underground mining

Primary lode deposits are usually worked by underground mining, access normally being gained by sinking vertical or inclined shafts, with horizontal tunnels driven from the main shafts so as to intersect the tin-bearing veins at right angles. Further tunnels are then made along the lodes so that the ore can be exposed. In mountainous country, tunnels may be driven directly in from the hillsides and the need for the more expensive vertical shafts is minimised.

In a typical underground tin mine, several lodes are worked simultaneously on different levels. Although groups of lodes are generally roughly parallel, they may branch, cross or join one another, or the veins may widen out or narrow. Furthermore, the distribution of cassiterite in the veins may be uneven. Consequently the exploration and development of an underground mine is always somewhat uncertain. A team of skilled geologists is employed to assist in assessing the economic viability of each working zone.

### Hydraulic mining – gravel pumping

Unconsolidated deposits of tin minerals on hilly or sloping ground, and small deposits on flat ground, are usually worked hydraulically by the method known as gravel pump mining, illustrated in Fig. 2. It is very widely practised in South East Asia, and throughout the world this method probably accounts for 40 per cent of all tin production.

In gravel pumping the tin-bearing material is broken up and washed away from the near-vertical faces of the pits by directing on to them high-pressure jets

Fig. 3 – Large tin mining dredge used to scoop up alluvial or sea-bed deposits of tin ore. Mineral concentration processes are carried out on board and the tailings are deposited over the stern of the vessel.

of water from movable nozzles or 'monitors'. This slurry of stones, gravel, sand and mud usually flows by gravity to a central sump area. The gravel pump itself, which is positioned above the sump, sucks up the slurry and pumps it to the treatment plant.

### Dredging

Dredging, by means of bucket or suction-cutter dredges, is an important method of mining tin ore, especially in South East Asia, where it is used for recovering unconsolidated ore from river beds and from beneath artificial lakes or paddocks and even to mine offshore deposits from the sea bed.

The dredges themselves (Fig. 3) are huge floating processing plants, which move across the water digging out tin-bearing sediments from the bed and elevating them to the mineral processing plant on board. After passing through a primary tin ore recovery plant, the waste material is discharged over the stern. Sizes of tin dredges vary, depending on such factors as the amount of material to be processed, the depth of operation and the amount of barren overburden above the tin-rich strata. Dredges range from those handling about 200 tonnes of material per hour from a depth of about 10 m to large plants which can excavate and process over 1000 tonnes per hour recovered from up to 50 m below the water level.

### Open-cut mining

Open-cut mining is a less widely employed method of mining for tin. The principal methods of working open-cut mines are by benches or with inclines and the ore is removed using mechanical shovels, drag-lines, excavators or even by manual labour.

Working by benches is used in mining large, flat, alluvial deposits up to 20 m deep; the maximum height of bench is limited by the danger of caving to about 7 to 8 m. Inclines are often used in the deeper mines up to 70 to 80 m deep.

Ore excavated from the working faces may be loaded into cars or trucks for transport to the primary treatment plant. Appropriate methods are used to stabilise the slopes after the tin-bearing material has been removed.

## 2.3 ORE TREATMENT

As mentioned earlier, the tin content of a typical ore body is very small, so that extensive pretreatment of the run-of-mine ore is needed in order to produce a high grade tin 'concentrate' that is suitable for smelting to obtain the metal.

Fortunately the specific gravity of the principal tin mineral, cassiterite, is much higher than that of the granite and other minerals with which it is normally associated, so that gravity separation techniques may be readily employed. Nevertheless, due to the leanness of the original ore and the high value of the tin, the mineral processing flow-sheets are frequently very complex and the re-circulating load within the plant may be many times the final output per hour.

In plants treating lode ore, and on the tin dredges, the first stage of separation is frequently by jigs. A jig is essentially a rectangular tank, full of water, with a horizontal metal screen across it near the top. The ore is fed on to one side of the screen and the water is forced up and down through it by a plunger or diaphragm in the lower part of the tank. The pulsing of the water opens up the bed of ore particles and the cassiterite and other heavy minerals work their way down and are withdrawn from beneath the screen. The lighter valueless material is carried over and away by the flow of the water. Several stages of jigging are required to produce a high grade of concentrate. Jigs treat ores in the range of 10 to 2.5 mm; finer sizes are treated by tabling.

A shaking table has a flat surface about 5 m long by 2 m wide and can be tilted slightly from the horizontal. The table is shaken with a differential motion in the direction of the long axis. The feed material in the form of a slurry with copious quantities of water is poured via a trough along the upper, long edge of the table and flows across the surface of the table. Most tables are equipped with longitudinal ridges or ripples to control the fluid flow.

The combination of the vibration and the flowing film of water causes the particles to become stratified, with the heavier particles of cassiterite being carried between the ripples to the short end of the table, whilst the lighter waste sands are washed down the table and off the long edge. Most tables produce three products, a concentrate, tailings, and a 'middlings' or mixed product, which is re-tabled. Tables usually treat material in the size range from about 2 mm down to 7 to 8 $\mu$m.

Tin ore recovered from gravel pump mining is usually treated first in a sluice box or 'palong'. This is a sloping wooden trough about 2 m wide by 1 m deep and 30 m long. As the sand and slurry from the gravel pump flow down the palong, the cassiterite and other heavy minerals settle to the bottom. From time to time the feed is diverted and the impure concentrate removed for further treatment by jigs and tables as described already.

Tin concentrates from the mineral treatment plant are dried and bagged ready for sale to the smelter. Concentrates from the Far Eastern alluvial deposits are very pure and may contain over 70 per cent tin. Concentrates from the complex lode ores are more difficult to purify and may contain as little as 20 to 25 per cent tin.


## 2.4 EXTRACTION AND REFINING

Tin concentrates obtained as described in the preceding pages are primarily cassiterite, the naturally occurring oxide of tin. To obtain tin metal from the concentrate it is smelted, by reduction with carbon at high temperature. The impure metal derived from primary smelting is purified by a series of refining operations until the required grade of commercial purity is obtained.

Before smelting, however, it is often necessary to treat the concentrates to remove the metal impurities they may contain. Generally speaking, the high-grade concentrates from alluvial deposits require less preparation than the lower-grade concentrates arising from lode mines.

Several methods are available for removing metal impurities from tin concentrates prior to smelting. Most are based on roasting, i.e. heating in air with or without the addition of chemicals to assist separation. During roasting, volatile constituents such as sulphur and arsenic are driven off and may be recovered. Roasting is sometimes followed by leaching with either water or acid solutions to remove impurities that have been rendered soluble by roasting. Other impurities may be made soluble by roasting with sodium carbonate or by a chloridising roast with sodium chloride.

Following this preliminary treatment, the dried concentrate is mixed with carbon, usually in the form of powdered anthracite, and with limestone to act as a flux.

### Tin smelting

The reduction of tin from its oxide to the metal is not difficult, as the reduction takes place fairly readily when tin oxide is heated to fairly high temperatures in the presence of reducing agents. The basic reaction is

$$SnO_2 + C \longrightarrow Sn + CO_2 \ .$$

However, in practice, tin smelting is complicated by the fact that tin oxide combines readily with silica to form tin silicates and therefore during the smelting of tin concentrates, which are usually contaminated with silicate minerals, a considerable portion of tin invariably ends up in the slag with the silicates. Because of the valuable nature of the metal, it is necessary to re-treat these slags to recover the tin from them.

The smelting of tin, therefore, consists of three stages: primary smelting to produce an impure metal and tin-rich slags; re-treatment of the first-run slags to recover the tin; and refining of the resultant metallic tin from the smelting processes.

One of three types of furnace is generally used for tin smelting: reverberatory furnaces, blast furnaces and electric furnaces. In a typical reverberatory furnace (the most widely used type), the charge, consisting of the tin concentrate mixed with anthracite and flux, sometimes mixed with small amounts of slag and refinery by-products, is heated at 1300 to 1400 °C for a period of some 10 to 15 h, with frequent mechanical stirring especially towards the end of the heat.

After a batch is smelted, it is tapped from the furnace and the reduced metal and slag are allowed to run into large settling pots. In these the heavier metal sinks and the slag overflows and is preferably dropped into water to granulate it to assist in the re-treatment step. The slag produced in the first-run smelting of tin concentrates invariably contains a high proportion of tin and must therefore be re-treated. Normally it is smelted in a similar, but separate, furnace, but the conditions differ somewhat. Higher temperatures, greater quantities of reducing agents and longer times are required. Slag smelting requires great expertise to remove the bulk of the tin and leave a second slag that can be discarded. A typical reverberatory furnace may take up to 15 tonnes of charge in each batch. After the furnace has been tapped, the next charge is introduced as soon as possible, before the furnace has time to cool.

Smelting in blast furnaces (the oldest method) or electric furnaces (the most modern) differs in detail but not in the basic principles from the method described.

The impure tin from the primary and slag smelting stages is cast into ingots or slabs for further refining to remove metallic impurities from the concentrate which are reduced with the tin and become alloyed with it. Two methods of refining are practised: heat treatment (or fire refining) and electrolytic refining.

**Fire refining**

The fire refining method of refining impure tin consists of one or both of the operations of 'liquating' and 'boiling'. When both are necessary, liquating is carried out first.

The object of liquating, or 'sweating', is to remove those impurities, alloys and compounds that have melting points appreciably higher than that of tin. It is carried out in a small, sloping hearth reverberatory furnace, with the slope of the hearth towards the tap-hole. The ingots or slabs of tin to be refined are placed on the upper end of the hearth and heated slowly to just above the melting point of tin (232 °C). Careful regulation of the temperature is necessary to ensure that the tin melts slowly and steadily, so that it trickles down the hearth and is caught in a vessel outside the furnace. After the first run tin is removed, the remaining dross is heated to a higher temperature and a second grade of sweated tin is produced and collected separately. This less-pure tin is returned to the liquating furnace for re-treatment. Finally the dross from the second liquation is roasted in the furnace and subsequently returned to the primary smelter as part of the furnace charge. The liquating process removes most of the iron, arsenic, copper and antimony from the smelted tin as intermetallic compounds of high melting point. Lead and bismuth, if present, are not removed by this process.

The first run of liquated tin is subjected to further refining by boiling or 'poling'. The metal, in a cast iron vessel (known as a kettle; see Fig. 4) containing 5 to 10 tonnes, is heated to a temperature considerably above its melting point, i.e. to about 300 °C, and is stirred continually. The metal is stirred or 'boiled' either by passing compressed air through it or by immersing in it poles or sticks of green wood, which decompose and give off gases which agitate the molten metal. By this vigorous stirring action, the impurities are brought to the surface, where they are oxidised to dross and skimmed off. If the tin contains fairly large amounts of impurities, boiling can take several hours. The dross skimmed from the boiling kettles is either re-liquated or re-smelted with the original concentrates.

Fig. 4 — Tin metal extracted from the oxidic ore by reduction with carbon is transferred to a large cast iron kettle for refining.

**Electrolytic refining**

Electrolytic refining is used in some smelters to upgrade the tin produced by thermal refining methods, more especially when the original concentrate was complex and hence the fire-refined tin may contain metallic impurities. Electrolytic refining is also frequently used in plants which recover secondary tin from tin-bearing scrap such as used bearings, solders, tinplate scrap, etc. As in the electrolytic refining of other metals, many impurities originally contained in the tin are removed by this process and a very pure grade of tin is produced.

In electrolytic refining, the impure tin is cast into flat sheet anodes of suitable size and suspended from bus bars in a tank of electrolyte. The cathodes, or starting sheets, are thin sheets cast from electrolytically refined tin. The electrolyte in tin refining is usually an acid solution based on sulphuric or hydrofluosilicic acid, together with addition agents to ensure a smooth, adherent deposit (see Chapter 8). For de-tinning and secondary recovery an alkaline electrolyte may be employed.

Electrolytic refining is a batch process and generally involves a large number of plating cells. When a low-voltage direct current of around 100 $A/m^2$ is applied between the anode and cathode, the anode dissolves and refined tin is deposited on the cathode. The process may take several hours and generally produces a tin of at least 99.99 per cent purity compared with 99.75 to 99.85 per cent purity generally obtained by fire refining. During electrolytic refining, the anode impurities settle out in the tanks as sludge and the electrolyte must be circulated and filtered to remove this residue.

Refined tin, whether obtained by fire refining or after electrolysis, is normally re-melted and cast into ingots usually weighing 50 kg. The brand name is generally cast into the ingot surface.

## 2.5  SECONDARY TIN – RECOVERY AND RECYCLING

All minerals constitute 'wasting assets' in the sense that, once mined, they are not replaced in the ground, at least not within a measurable time span. Metals, however, unlike the fossil fuels, can be recycled, so that they do not represent a total loss but a recoverable resource.

In the case of tin, recycling can be broadly classified into three categories — recovery of tin metal (secondary tin), recycling of high-tin alloys, and re-use of low-tin alloys. The published statistics of secondary tin production and use do not as a general rule include the two latter categories and hence give a somewhat misleading impression as to the proportion of tin that is conserved by recycling.

Secondary tin may be recovered from high-tin alloys such as pewter, white-metal bearing alloys and solder by fire-refining methods similar to those used for primary tin production described earlier in this chapter. The major difference is that the impurities to be removed are the result of deliberately added alloying elements and are therefore usually present in greater proportion that in first-run primary smelted tin. The principal alloying elements which may need to be removed when refining scrap and residues are antimony, arsenic, copper, iron, lead and nickel, although others may also be present.

An important source of secondary tin is the process scrap generated by the can-making industry, for example, when stamping circles from strip or sheet tinplate, and scrap arising from tinplate manufacture. This material is collected, usually as loose, unbaled scrap, and sent for recovery of the tin and the steel. The usual process is for the tin to be dissolved from the steel in hot alkaline solutions and for the tin to be recovered electrolytically, as for electro-refining described earlier.

In addition to the tinplate process scrap which is sent for de-tinning, there is considerable interest at the present time in collecting and reclaiming used tinplate containers. A large number of municipal authorities in several countries have installed magnetic treatment plants in their refuse collection depots with the purpose of removing tinplate cans for subsequent treatment. It has been demonstrated that it is technically possible to recover the tin and the steel from reclaimed tinplate cans separately. The question of whether it is an economically attractive proposition depends on a number of factors, only some of which involve the relative prices of the materials.

Electrolytic refining for the production of secondary tin from high-tin alloy scrap is also practised, as described earlier in this chapter.

To make a secondary use of the tin content of tin-containing scrap it is often not necessary to reclaim high-purity tin, and indeed it is usually not economically advantageous to separate the scrap alloys into their constituent elements. Many producers of tin alloys, such as solders or whitemetals, employ high-tin scrap as an important source of their raw materials.

The procedure is first to segregate the incoming scrap carefully, with due regard to the analysis of the materials. The scrap is then melted down in suitable vessels and the bath analysed for the major elements. Suitable additions are made of alloying elements needed to achieve the required specification of the new alloy and any necessary refining is carried out.

Whilst high-tin scrap such as whitemetals, solders or pewter, may be treated either as a source of high-grade secondary tin or as the raw materials for further alloy production, it is very rare indeed for tin-bronzes to be used as a source of secondary tin. Bronze scrap is almost invariably remelted and used as the basis for new bronze alloys by judicious addition of suitable alloying elements. Similarly, tinned copper wire scrap arising from use in the electrical industry is not normally de-tinned but is much more likely to be used as a source of both tin and copper for bronze production.

## FURTHER READING

1. Mantell, C. L., *Tin*, Hafner Publishing Company, New York, 1970.

2. Wright, P. A., *Extractive Metallurgy of Tin,* Elsevier, Amsterdam, 1983.
3. Thews, E. R. *Metallurgy of Whitemetal Scrap and Residues,* Chapman and Hall, 1930.
4. *Proceedings of Technical Conferences on Tin,* International Tin Council, London, 1967, 1969, 1974.
5. Gmelin, L., *Handbuch der Anorganischen Chemie – Zinn,* Teil A, Verlag Chemie, Weinheim, 1971.
6. *The Occurrence and Production of Tin,* International Tin Research Institute Publication No. 448.
7. Read, H. H., *Rutley's Elements of Mineralogy,* Allen & Unwin, London, 1970.
8. Denyer, J. E., The Production of Tin, *Proceedings of the Conference on Tin Consumption,* International Tin Council, London, p. 47, 1972.
9. *Materials Survey – Tin,* US Department of Commerce, 1953.
10. Fawns, S., *Tin Deposits of the World,* 3rd edn., The Mining Journal, London, 1912.

# 3

# Tin Alloy Systems

Tin forms binary systems with many other metals and non-metals and in a study of the properties and applications of tin-containing materials, it is frequently helpful to metallurgists and engineers to have available the equilibrium phase diagram for the system under consideration. This is true whether the alloy system is for commercially important alloys of low tin content, such as titanium and zirconium basis materials, or is related to tin-rich alloys such as whitemetal bearing alloys. However, in the more important industrial applications of tin-containing materials some equilibrium phase diagrams have been included and described in Chapters 4, 5 and 6 concerning the technology of soft solders, bearing metals and other alloys such as bronzes. In this chapter therefore, are included diagrams concerning the less important binary alloy systems. Additionally, some diagrams are included in order to clarify metallurgical reactions occurring, for example, at interfaces during the soft soldering of goldplated surfaces and the flow melting process used in the manufacture of tinplate (see Chapter 9).

Hence, in this chapter are presented a number of diagrams which are generally of incidental importance in the industrial applications of tin and its alloys but are of scientific value, and cross reference is made to the chapters describing the major applicational areas where appropriate. For brevity, only that information which is of importance to the particular process, alloys or reactions occurring has been included. The equilibrium phase diagrams are given in the alphabetical order of the second element. For completeness some notes are added for a few selected ternary and quaternary tin-alloy systems which are relevant to commercially used alloys.

Some selected photomicrographs of some of the alloys are either included in this chapter or are referred to in other chapters concerned with specific applications of tin and tin alloys. Since tin alloys are relatively soft, both these materials and tin coatings on harder substrates require rather special metallographic preparation techniques and some notes on these are given in Appendix 1 to be used as a guide.

## 3.1 INTERMETALLIC COMPOUNDS OF TIN

In metallurgical terms, a particular characteristic of tin is that it readily forms discrete intermetallic compounds in alloy systems with other metals. In a few cases only (for example indium, cadmium and mercury) are tin-rich intermediate phases formed, having a limited range of composition. The compounds formed in alloys containing tin are often a source of additional strengthening or hardening as, for example, is the case with the hard $\delta$ phase present in the higher tin content cast tin bronzes, which imparts wear resistance. The only tin-containing intermetallic compound which has achieved commercial importance in its own right is niobium–tin, $Nb_3Sn$, which is one of the widely used superconducting materials. The ternary compound PbSnTe, also has industrial use as a semi-conductor for electronic devices.

In several binary systems of tin with other metals, wide composition range solid solutions are formed. Many of these exhibit a large reduction in solid solubility as the temperature is reduced towards room temperature, causing precipitation of a second phase from solution, but in no case does this result in any significant strengthening, such as occurs in many other metallurgical age-hardening systems.

## 3.2 TIN–ALUMINIUM

A simple binary eutectic forms at the tin-rich end of this system and the liquidus line increases steeply with temperature as the aluminium concentration in the

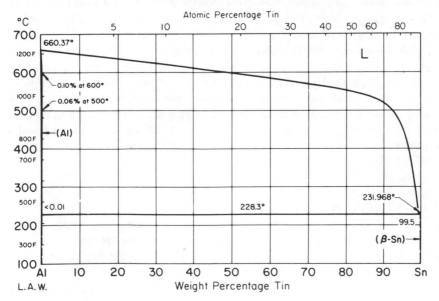

Fig. 5 – Tin–aluminium diagram. (Courtesy: *Amer. Soc. Metals Handbook.*)

alloy is raised. Alloys richer in aluminium than the eutectic composition precipitate virtually pure aluminium as the primary phase on cooling. It will be seen that at only 100 °C below the melting point of aluminium, the solubility of this element in molten tin is still only about 15 per cent by weight. There is complete miscibility of the liquid metals. The solubility of tin in solid aluminium just below the melting point is in the region of 0.1 per cent and much lower (e.g. *ca* 0.01 per cent) at 100–200 °C. Microstructures of aluminium–20 per cent tin alloys are seen in Figs. 65 and 66.

The diagram is of importance in the aluminium–tin bearing alloys (Chapter 6) where the usual compositions used commercially are in the range 20 to 40 per cent tin. In this case 1 per cent or so of copper is also present but this is completely in solid solution in the aluminium and does not significantly affect the equilibrium diagram. The solid solubility of aluminium in tin has not been determined but is such that only about 0.001 per cent aluminium in pure tin foil is as a second phase which in a humid atmosphere causes intergranular cracking and disintegration of the foil in certain cases.

### 3.3 TIN–ANTIMONY

The equilibrium diagram for this system is given in Fig. 41 (Chapter 5) since the primary interest is the use of one particular alloy as a soft solder. A point to be noted is the significant decrease in solubility of antimony in tin from the liquidus to room temperature which may cause precipitation of the $\beta$ phase (SbSn). In alloys containing more than about 7 per cent antimony, for example, used in certain bearing metal compositions (Chapter 6), under normal chill casting conditions a separation of $\beta$ occurs to produce cuboids of a phase of composition nominally 50 per cent antimony–50 per cent tin but with a range of about 41 to 57 per cent tin at room temperature. Addition of small amounts of copper in tin–antimony alloys, for example in pewter or some bearing metals, gives rise to a diagram similar to that of the binary tin–copper alloys with regard to liquidus line. Die casting tin-based alloys may contain 20–40 per cent antimony and a typical structure of such an alloy is seen in Fig. 20 (Chapter 4).

### 3.4 TIN–BISMUTH

This is a classical binary eutectic system with a eutectic temperature of just under 140 °C which gives rise to the current and increasing application of alloys of around this composition as low-temperature soft-soldering alloys. The solubility of bismuth in tin at the eutectic temperature is nearly 20 per cent but falls to virtually zero at room temperature. The solubility of tin in bismuth is probably less than 1 per cent.

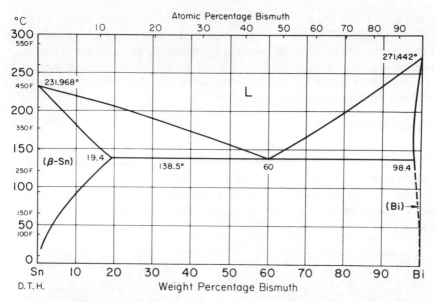

Fig. 6 – Tin–bismuth diagram. (Courtesy: *Amer. Soc. Metals Handbook.*)

## 3.5 TIN–CADMIUM

A binary eutectic occurs in this system at a temperature of 177 °C and 68 per cent tin which gives rise to a possible use for such alloys as low-melting-point soft solders. However, it is more usual to introduce lead into such alloys and

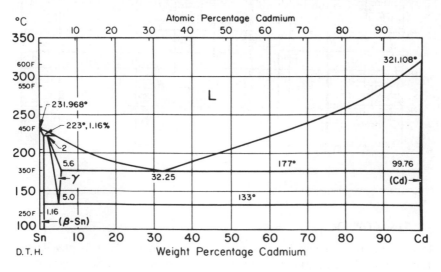

Fig. 7 – Tin–cadmium diagram. (Courtesy: *Amer. Soc. Metals Handbook.*)

make use of the ternary eutectic of lead, cadmium and tin (*q.v.*) melting at 145 °C (Chapter 5). The remainder of the binary diagram is not of commercial interest although an alloy 70 per cent cadmium—30 per cent tin is sometimes used as a solder in electronics because of its particular electrical properties, whereby the thermal e.m.f. developed between this alloy and the basis materials in a soldered joint is virtually zero.

## 3.6 TIN—COBALT

The diagram, Fig. 8, shows that there are a high and a low melting point eutectic in the binary system and several intermetallic phases are formed. Of commercial interest is the $\gamma$ or $\gamma'$ phase which may be electrodeposited as a hard and corrosion-resistant coating for decorative or wear applications (Chapter 8), and is some-times used as an alternative to tin—nickel alloy electrodeposits (see also that system).

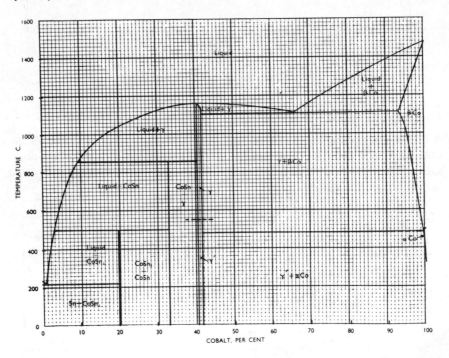

Fig. 8 – Tin—cobalt diagram.

## 3.7 TIN—COPPER

The equilibrium diagram for this system is given in Fig. 25 (Chapter 4) and is discussed in relation to tin bronzes where the tin content may vary approximately up to levels of 15 per cent for engineering applications or over 20 per cent for cast bells (see also *Copper and Its Alloys* in this series). Microstructures are seen in Figs. 26—32. The eutectic formed at about 0.7 per cent copper in tin is the composition of a material sometimes used for the suitability of its forming properties in the production of impact extruded tin tubes for medical purposes. Of interest are the two phases $\eta$ and $\epsilon$ corresponding to the compositions $Cu_6Sn_5$ and $Cu_3Sn$ which inevitably form when liquid tin is brought into contact with copper, as in the case of hot dip tinning of copper articles or in the formation of soldered joints to copper (Figs. 49 and 112). These compounds also form by solid-state diffusion at the interface in tin-plated copper subjected to long term or elevated temperature storage.

## 3.8 TIN–GOLD

The phase diagram shown in Fig. 9 indicates the formation of a series of inter-metallic phases and two eutectics. The one of higher melting point, containing

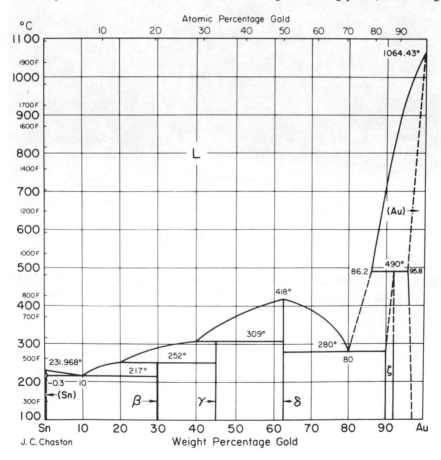

Fig. 9 – Tin–gold diagram. (Courtesy: *Amer. Soc. Metals Handbook.*)

80 per cent gold—20 per cent tin, is sometimes used as a special joining alloy in the microelectronics industry. Of interest, is the formation of $AuSn_4$ and perhaps the other phases during the soft soldering of gold-plated surfaces in which instance problems arise due to the embrittlement of the solder joint by the presence of these phases. Figure 10 illustrates the result of leaching of a thick

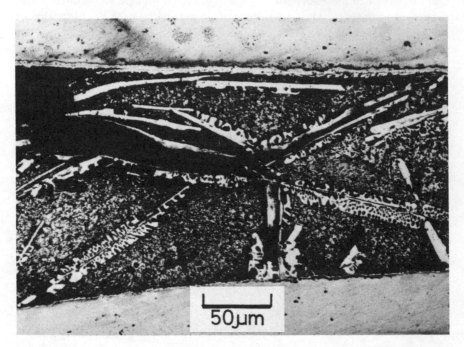

Fig. 10 — Massive primary dendrites of $AuSn_4$ in a tin lead soldered joint resulting from soldering on a thick gold plating.

gold-plated surface into the solder to produce large $AuSn_4$ dendritic crystals which when present in sufficient quantity cause brittle joints. Similar phenomena do not apply when silver is soft soldered.

## 3.9 TIN–INDIUM

This is a eutectic system, Fig. 11, with approximately equal weight percentages of the elements forming the eutectic composition which melts at a temperature of 120 °C. Alloys of this composition are used sometimes as low-melting soft-solder alloys especially because the presence of indium seems to confer the special property of wetting and bonding to glass or glazed surfaces. There is also commercial interest in the ternary system of tin—lead—indium as a solder alloy.

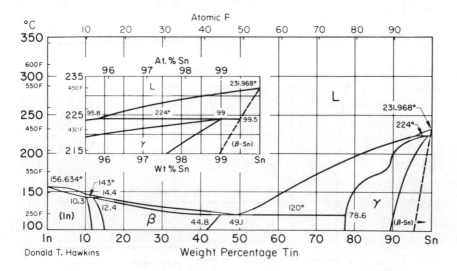

Fig. 11 – Tin—indium diagram. (Courtesy: *Amer. Soc. Metals Handbook.*)

## 3.10 TIN–IRON

The equilibrium diagram, Fig. 12, shows the phases present in this binary system and it will be noticed that a γ-loop is present at the iron-rich end of the diagram. The solubility of tin in α-iron is nearly 10 per cent at room temperature, increasing to about 19 per cent at 860 °C. Two intermetallic phases are in equilibrium at room temperature namely $FeSn_2$ and FeSn. However, in the process of coating steel by dipping in liquid tin (Fig. 90) or during the manufacture of tinplate in which a plated tin coating is momentarily melted (Chapter 9), only the presence of the compound $FeSn_2$ is observed (Fig. 86). The presence of carbon in iron alters the position and size of the γ-loop at the iron-rich end of the diagram as shown in Fig. 13 (p. 43). This is of importance when considering the influence of tin on the sintering temperature used for powder compacts containing iron and tin (Chapter 4).

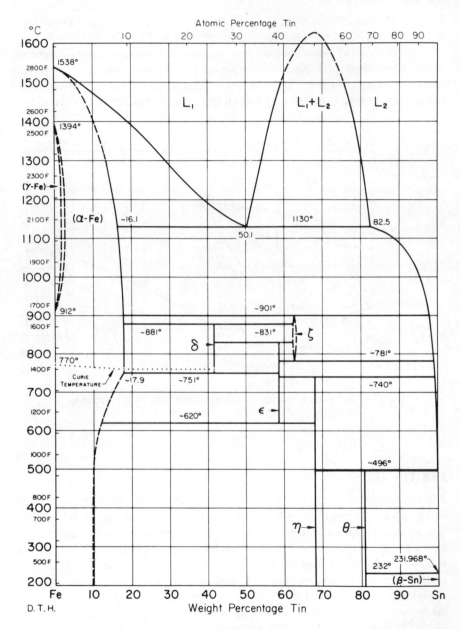

Fig. 12 — Tin—iron diagram. (Courtesy: *Amer. Soc. Metals Handbook.*)

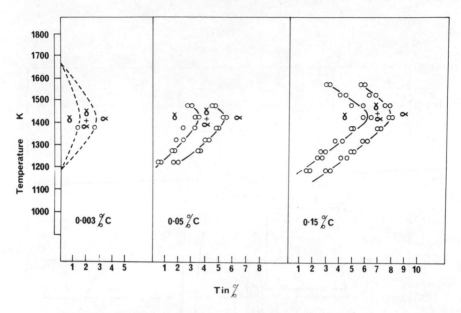

Fig. 13 – Effect of carbon on γ-loop in the Fe–Sn system.

## 3.11 TIN–LEAD

This alloy system is of particular importance in relation to soft solders and their uses, the reference should be made to Fig. 39 (Chapter 5) for details of the phase diagram. It is a classical eutectic system with solid solutions at each end of the diagram. In general, alloys of composition throughout the whole range of mixtures are of interest for use as soft solders, and the liquidus–solidus gap will vary with the composition selected. Microstructures of tin–lead alloys are given in Fig. 40.

## 3.12 TIN–NICKEL

The phase diagram, Fig. 14, resembles that for tin and cobalt, and three inter-metallic phases are formed. These may be observed as thin layers of intermetallic compound when molten tin is brought in contact with nickel, as in the formation of soldered joints on nickel (cf. Fig. 113). The electroplated tin–nickel alloy of commercial interest (Chapter 8) has a composition corresponding to about 65 per cent tin–35 per cent nickel or NiSn and is a single metastable phase when electrodeposited but decomposes to form the compounds according to the phase diagram when heated to temperatures above 500 °C.

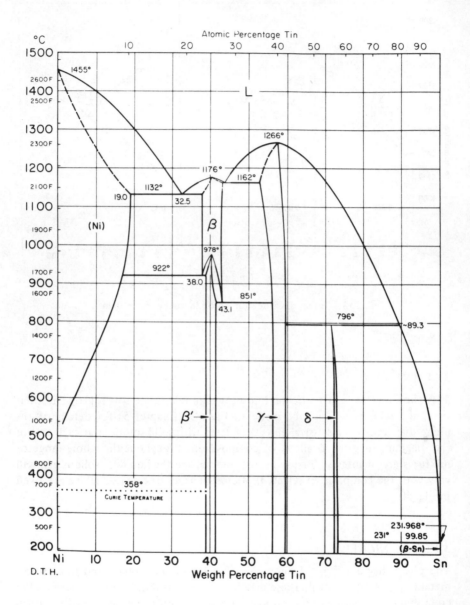

Fig. 14 – Tin–nickel diagram. (Courtesy: *Amer. Soc. Metals Handbook.*)

## 3.13 TIN−NIOBIUM

The feature of interest in this binary system is the formation of the phase $Nb_3Sn(\epsilon)$ which is one of the phases having a high superconducting temperature and is therefore currently used commercially in the manufacture of superconducting electromagnets. Figure 15.

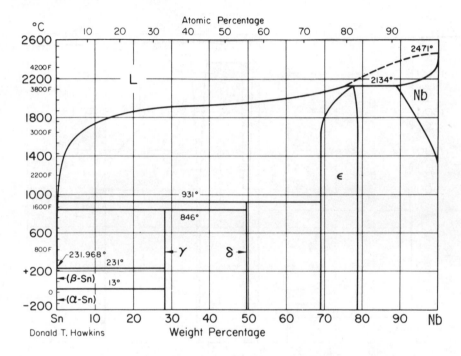

Fig. 15 − Tin−niobium diagram. (Courtesy: *Amer. Soc. Metals Handbook.*)

## 3.14 TIN−SILVER

The phase diagram is given in Fig. 42 (Chapter 5), since the alloys of interest are those rich in tin used as soft solders and are based on the eutectic composition. During the soldering of silver coated surfaces the intermetallic compound $Ag_3Sn$ will form, as seen from the diagram. The solubility of tin in silver is about 10 per cent at 200 °C. As noted in the chapter on soft solder alloys, alloys of tin, silver and lead are also frequently used (see notes on this ternary system).

## 3.15 TIN–TITANIUM

The phase diagram shows a relatively complex system with high melting point phases corresponding to $Ti_3Sn(\zeta)$, $Ti_2Sn(\epsilon)$, $Ti_5Sn_3(\delta)$ and $Ti_6Sn(\gamma)$. Alloys of titanium particularly of use in the aerospace industries sometimes contain tin to promote formation of the $\alpha$ phase (Chapter 11). The majority of such alloys have a tin content of a few per cent only.

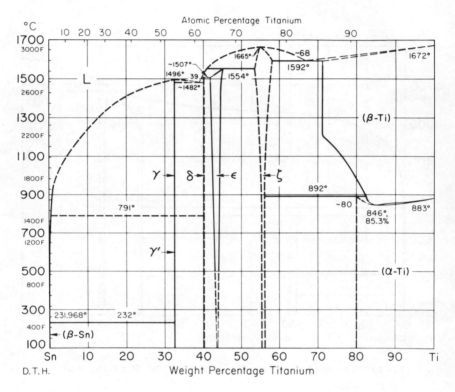

Fig. 16 – Tin–titanium diagram. (Courtesy: *Amer. Soc. Metals Handbook.*)

## 3.16 TIN–ZINC

The diagram (Fig. 45) is given in connection with the use of certain alloys as soft solders for aluminium (Chapter 5). Specifically, these alloys contain 10–30 per cent zinc and therefore consist of a primary zinc phase (the solubility of tin in zinc is less than 0.1 per cent) in a matrix of tin–zinc eutectic which has a melting point of 199 °C. The only other commercial application for alloys of these two elements is an electroplated coating containing around 20–25 per cent zinc which protects steel in a similar manner to pure zinc coatings (Chapter 8).

However, the microstructure of such electrodeposits appears to consist of small particles of zinc in a pure tin matrix.

## 3.17 TERNARY ALLOYS

### Tin—Antimony—Copper

Alloys of commercial interest are pewter (Chapter 4) and bearing metals (Chapter 6). Pewter and some whitemetal alloys have structures consisting of a dispersion of $Cu_6Sn_5$ in a solid solution of antimony in tin where Sb < 7—8 per cent but stronger bearing metals may have a higher antimony content and show the presence of SbSn as cuboids in the microstructure (Fig. 59). The relevant corner of the ternary equilibrium diagram is given in Fig. 58 (Chapter 6).

### Tin—Antimony—Lead

This system is of relevance to those tin—lead alloy soft solders containing a few per cent of antimony (Chapter 5) but for practical purposes the structures are as for the binary tin—lead alloys and the equilibrium diagram for that system is affected only slightly by the presence of antimony up to, say, 3 per cent.

### Tin—Cadmium—Lead

The importance of this system is the presence of a ternary eutectic of composition 50 per cent tin—32 per cent lead—18 per cent cadmium which melts at 145 °C and is useful as a low-melting-point soft solder (Chapter 5).

### Tin—Lead—Silver

Alloys covered by this system are of two types, firstly the binary tin—lead soft solder alloys doped with a small amount of silver to prevent leaching of silver-coated surfaces during soldering (e.g. 62 per cent tin—36 per cent lead—2 per cent silver) and secondly the lead-rich solders of higher melting point and mechanical strength (e.g. Pb— 1—5 per cent tin—1.5 per cent silver) (see Chapter 5). The microstructure of the former alloy is identical to that of the binary tin—lead alloys whereas the lead-rich alloys have a structure of the form seen in Fig. 44.

## 3.18 QUATERNARY ALLOYS

### Tin—Copper—Nickel/Zinc

For details of these copper-rich alloys see the book *Copper and Its Alloys* in this series.

### Tin—Antimony—Copper—Lead

Certain low-grade bearing metals are based on this system and are lead-rich in composition. They therefore generally have a matrix of lead—antimony solid solution with intermetallic phases such as SbSn and $Cu_6Sn_5$ present and perhaps ternary eutectic (see Fig. 61, Chapter 6).

### Tin–Silver–Copper–Mercury

These alloys form the traditional basis for dental amalgams which rely on reaction between the liquid mercury and the other solid constituents to form various intermetallic phases (mostly $Ag_3Sn$) of the desired hardness, wear- and corrosion-resistance (Chapter 11).

### Tin–Lead–Cadmium–Bismuth

The structure of alloys in this system has not been fully investigated and these alloys are used because of their very low melting point as fusible alloys (Chapter 4) such as Wood's metal which is liquid at 70 °C.

### FURTHER READING

1. *Metallography of Tin and Tin Alloys,* International Tin Research Institute Publication No. 580.
2. *Metals Handbook,* 8th Edn., Vol. 8, American Society for Metals, Ohio, 1973.
3. *Equilibrium Data for Tin Alloys,* International Tin Research Institute Publication No. 165.

# 4

# Tin and Tin Alloy Applications

As indicated in Chapter 1, tin has unique properties of low melting point, softness allied to lubricity, non-toxicity and an ability to alloy with most of the common metals. These are the properties which are exploited in its numerous metallurgical applications. The major use of tin in alloys is as soft solder, accounting for about one quarter of the present tonnage use of primary tin, and for convenience Chapter 5 is devoted to this important subject. Tin is also well known for its use in bearing metals and it was considered desirable to devote a separate section (Chapter 6) to this subject. The present chapter, therefore, describes and discusses only the applications of pure tin and the tin alloys not covered by the succeeding two chapters.

The alloys of tin may be broadly classified into high- and low-tin alloys. In the former, tin is the major, or at least a substantial, constituent. An example of a tin-rich alloy is pewter (and, of course, others are solders and whitemetal), while fusible alloys and type metal often also contain significant quantities of tin. The classical examples of lower tin content alloys are those based on copper − the tin bronzes and gunmetals. More recently, substantial tonnages of tin are finding use in the cast-iron and powder-metallurgy industries.

## 4.1 PURE TIN

The low hardness and consequent low mechanical strength of pure tin make it unsuitable for use as a bulk material of construction unless strengthened by hardening alloying elements. In its unalloyed state, therefore, the uses of tin are confined to those where its excellent malleability, non-toxicity or other special properties are required.

*Tin foil* is produced by cold rolling cast ingots progressively to as thin as 0.004 mm. A principal use is in certain high-grade dry electrical capacitors in which tin foil is interleaved with a paper dielectric layer. Tin foil is also used to a limited extent for wrapping chocolates and dairy produce, because it is non-corrosive and is devoid of springiness. This use, however, has largely been replaced by aluminium. Tin foil inserts are used in the caps of whisky bottles to prevent contact with the cork seal. Tin foil may also be used as pre-forms in special soldering operations where pure tin rather than tin—lead alloys must be used. Tin sheet impregnated with diamond dust is occasionally used as a surface for grinding or lapping.

*Tubes and pipes* are extruded from solid, cast billets and were formerly employed in many breweries for conveying beer from the cellar to the bar. A cooling coil in the tin pipe was usually immersed in brine to keep the beer at a desirable temperature.

*Tin wire* is extruded and drawn by normal techniques and is used for certain types of electrical fuses instead of copper wire.

*Collapsible tubes* are formed by impact extrusion and tin tubes are used now chiefly for dispensing medical products. Some tubes contain a built-in hypodermic needle for self-injection of pain-killing drugs on the battlefield. The valuable assets of tin in this application are non-toxicity and resistance to corrosion by the medical preparations packaged. A further advantage is its lack of elastic recovery, so the tube remains collapsed and can be completely emptied. Tin-coated lead tubes are still used to some extent for toothpastes, sauces, artists' paints, etc., but for these applications aluminium or plastics are being substituted.

*Molten tin* in bulk form provides the optically flat surface on which float glass is produced (Fig. 17). In this process, molten glass, at a temperature of

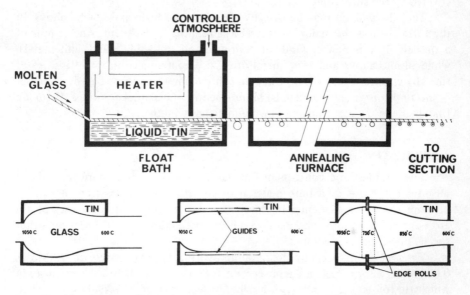

Fig. 17 — Diagram illustrating the principle of continuous sheet glass production by the 'float glass' method employing a large bath of molten tin to impart a flat lower surface to the semi-liquid sheet. (Courtesy: Pilkington Bros. Ltd.)

about 1000 °C, is poured directly from the furnace on to the surface of a large bath of molten tin under a controlled atmosphere to prevent oxidation. The molten glass, which has a specific gravity of 2.4, floats on the surface of the tin (specific gravity 6.5) and the glass solidifies whilst being supported by the molten tin. Because there is no contact with solid support whilst they are still hot, both surfaces of the glass are so smooth that they require no further polishing, unlike plate glass produced by earlier methods.

Molten tin is an ideal medium for this process; in addition to its special thermal properties there is little chemical interaction between the tin and the glass. The molten metal is also a good conductor of heat and hence equalises the temperature across the width of the glass. Furthermore, molten glass does not spread indefinitely over the surface of clean, molten tin, so that it is possible to control the thickness of the glass sheet obtained, by control of the operating parameters.

The normal range of glass thicknesses is from 3 mm up to around 15 mm. Typical applications of glass produced by this process (which can include safety glass) are for shop windows, mirrors, and windows and windscreens for the automobile industry.

## 4.2 PEWTER

Pewter is an alloy containing usually over 90 per cent tin and is widely used for utensils, such as tankards and goblets, or decorative items like plates and candlesticks or costume jewellery. The variety of bright or oxidised finishes available adds to the metal's attractiveness and therefore pewter has to some extent replaced the traditional silverware which was formerly part of the domestic scene.

Pewter is known to have been used extensively in Roman times and it is reported that Pliny, writing in the first century AD, stated that a tin vessel improved the  taste of wine. Originally the term 'pewter' was applied to any metal with a high proportion of tin, particularly a tin-lead alloy. The history of pewter can be traced, mainly from ecclesiastical artefacts, up to the fourteenth century when pewter began to replace wooden and pottery items for tableware and other household purposes. In 1348 the Worshipful Company of Pewterers was established in London to protect the craft secrets and to maintain high production standards. In the eighteenth century a new version of pewter known as Britannia metal was developed. This had a bright finish and contained a small amount of antimony but no lead. Britannia metal was harder than the other 'common pewter' and since it contained no lead it did not tarnish with age. Modern pewter, as illustrated in Fig. 18, is similar to Britannia metal as it is not alloyed with lead and is composed of about 92 per cent tin with normally about 6 to 7 per cent antimony and 1 to 2 per cent copper. Table 8 shows the compositions covered in BS 5140. In commercial alloys some bismuth, silver or other elements may also be present.

Fig. 18 — Examples of modern cast pewterware with a satin finish.

**Table 8**

Chemical composition of pewter (BS 5140)

| Alloy | Tin | Antimony | | Copper | | Lead | Cadmium |
|---|---|---|---|---|---|---|---|
| | | min. (%) | max. (%) | min. (%) | max. (%) | max. (%) | max. (%) |
| A | Balance, but not less than 91% | 5.0 | 7.0 | 1.0 | 2.5 | 0.5 | 0.05 |
| B | Balance, but not less than 93% | 3.0 | 5.0 | 1.0 | 2.5 | 0.5 | 0.05 |

**Cast Pewterware**

The low melting point and high fluidity of pewter allow the casting of very intricate designs and of thin-walled vessels. Traditionally, pewter articles were cast and the more expensive heavy items are still produced by this method, often requiring considerable skill. Reproduction 'antique' pewterware is very popular; these items may often be cast in the original and historic moulds. The general factors in melting and casting are dealt with in Chapter 7.

Occasionally large production runs may be made using sand moulds, in which case the procedure is virtually identical to normal sand-casting procedures used for other non-ferrous alloys. Such castings require machining before the final polishing operation.

Two modern, mechanised, methods of producing pewter castings are pressure die casting and centrifugal casting. Both processes are described in Chapter 7. Rubber-mould centrifugal casting is widely used for producing pewter jewellery, buckles, medallions etc.

### Spun Pewterware

One of the major ways in which modern pewter production varies from that of the past is in the increasing proportion of pewterware which is produced from rolled sheet by spinning or deep drawing. The availability of sheet pewter of uniform quality and in a range of thicknesses has done much to increase the standard of control and to permit the introduction of some mechanisation into the manufacture of pewterware by these techniques.

Pewter has extremely high ductility so that it can be stretched, compressed, hammered and bent into any desired shape. Unlike most metals, it does not work-harden and in fact it softens during deformation so that annealing is unnecessary. Advantage is taken of these properties in producing pewter items by spinning. The technique of spinning pewter is described in Chapter 7.

Deep-drawing techniques may also be applied to pewter sheet to form an initial plain cup which is then spun to its final contour.

Both cast and spun pewter items may be made in sections; the parts are then soldered together using a metal very similar in composition to the parts to be joined to ensure a good colour match. Great skill is required to produce perfect joints which are rendered invisible by the final buffing and finishing processes. For joints exposed to food, cadmium-free solder should be used such as 50 per cent tin − 40 per cent lead − 10 per cent bismuth, or the pewter alloy itself. Fixtures such as handles, spouts and hinges are also attached by soldering and for these 60 per cent tin − 40 per cent lead solder is used.

### Handles and Fittings

Handles for pewterware are usually cast. However to save weight they are usually not solid but are slush-cast in metal moulds (see Chapter 7). Some other fittings such as spouts may also be produced in this manner or, alternatively, cut from rolled sheet.

### Surface finishing

Cast and spun items are polished or buffed using rotating mops and different polishing media to produce surfaces ranging from highly reflective to a soft 'satin' texture. The dull grey finish of mediaeval pewter, characteristic of high-lead alloys, can be reproduced on modern lead-free pewter items by suitable chemical treatment. Additionally, other chemical treatments are available which produce black, coloured or textured surfaces. If an embossed surface is so treated, buffing can be used to relieve the high spots.

### Mechanical properties

Pewter alloys are used mostly because of their ease of fabrication into the required shape and the final strength properties are normally of little importance. However, in designing pewterware it should be noted that the inherent low strength limits the minimum section thickness that can be used, especially for elevated temperature service such as in teapots and water jugs. Cast pewterware is mechanically stronger than items fabricated from sheet since tin alloys, including pewter, work-soften during rolling or spinning. However, if about 2 per cent bismuth or 0.1 per cent silver is present in the tin—antimony—copper alloy and the fabricated material is heat-treated at about 150 °C, it develops a hardness value somewhat similar to that of cast pewter (Table 9).

Table 9

Hardness of pewter alloys after working and heat treatment

| Nominal composition (%) | | | | Vickers Hardness Value HV | | | % Recovery |
|---|---|---|---|---|---|---|---|
| Sn | Sb | Cu | Others | As cast | Rolled 90% | Rolled + 1 h/200 °C | |
| Bal. | 6.0 | 1.5 | — | 23 | 13 | 19 | 83 |
| Bal. | 6.0 | 1.5 | 0.1 Ag | 26 | 13 | 24 | 92 |
| Bal. | 6.0 | 1.5 | 2.0 Bi | 30 | 15 | 28 | 93 |

### Applications of Pewter

Pewter is largely used for domestic decorative items such as candlesticks and plaques or for drinking vessels like tankards and goblets. A very wide variety of such articles is available, some exhibiting modern art forms characteristic of their country of origin, others copying historical articles of pewterware. Similar alloys are used for centrifugal casting of figures such as knights, soldiers or jewellery in rubber moulds (see Chapter 7). It is of interest that a recent application of rolled pewter sheets is in the making of organ pipes to be used at the front of the instrument because of the higher tarnish resistance than the tin—lead alloy which is the traditional material (Fig. 19).

Fig. 19 – Modern organ pipes may be manufactured from rolled pewter sheet because of its resistance to tarnishing.

## 4.3 DIE-CASTINGS

Tin-base alloys were the first materials to be die-cast because their low melting point and extreme fluidity favour the easy production of sound castings of intricate design or pattern and without damage to the moulds. The technique was later extended to other metals of higher melting point and greater strength such as aluminium and zinc alloys (see for example the book *Zinc and its Alloys* in this series) until, in terms of the total number of die-castings made, the proportion made in tin alloys is now very small.

Tin-base alloys can be cast with greater dimensional accuracy than the other materials because of the low working temperature which minimises both distortion of dies and thermal shrinkage of the casting in cooling from the solidification temperature. It is possible to cast small parts with a maximum dimensional variation of ± 0.02 per cent and with very thin sections such as 0.5 mm.

Tin-base die-casting alloys generally resist corrosion sufficiently well for most purposes, but should the need arise, as for decorative purposes, they can be electroplated without difficulty. Corrosion resistance is important in some applications as, for example, in the slide valves of gas meters. In this case the bearing quality and wear resistance in unlubricated locations are important attributes.

### Composition of die-casting alloys

Tin-base alloys of such widely different compositions and properties have been die-cast that it is apparent that within reasonable limits any tin-base alloy can be cast successfully. The choice of alloy therefore depends on particular requirements such as mechanical properties, wear resistance or allowable cost rather than on any consideration of casting behaviour. However, in order to avoid high working temperatures, there must be a limit to those alloying elements used which have a low solubility in liquid tin. For example, in alloys with as much as 8 per cent copper, the liquidus temperature will exceed 400 °C and this would give rise to the risk of the molten alloy attacking either the die or the working parts of the casting machine.

Antimony is the principal addition to tin for die-casting purposes but small percentages of other elements such as copper and lead may be added. A typical alloy would be 70 per cent tin–30 per cent antimony. The structure of this would consist of primary SbSn crystals with a small proportion of tin-rich solid solution containing some antimony (Fig. 20).

The equilibrium diagram for this system is seen in Fig. 41 (Chapter 5). The solidification range of the alloys containing lead is somewhat longer than that of lead-free alloys and this may cause difficulty in casting, and the thermal contraction on cooling to room temperature is less for the lead-containing alloys. Tin–lead solders are also used for die-casting.

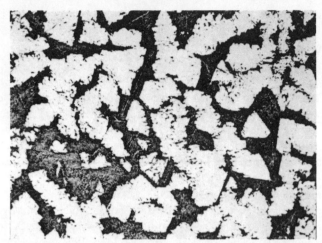

Fig. 20 – Microstructure of 70% tin, 30% antimony die-casting alloy with primary SbSn and tin–antimony solid solution. (Etched in 2% Nital.) Magnification ×100.)

For engineering purposes the variety of alloy compositions used is wide but, in general, alloys of higher lead content would be used where low frictional requirements are important, while for higher strength and wear resistance a higher antimony content is desirable because of the greater proportion of the hard intermetallic SbSn in the structure. Table 10 shows some typical alloy compositions and their applications.

**Table 10**

Some tin-containing die-casting alloys

| Composition (%) (balance Sn) | | | Typical uses |
|---|---|---|---|
| Sb | Cu | Pb | |
| 30–40 | – | – | Gas meter slides and grids |
| 25 | – | 25 | |
| 20 | 5 | – | Mechanical counter wheels |
| 10–17 | 2–8 | 4–12 | Electrical instrument frames and cases |
| 11 | – | 86 | Counterbalance weights |
| 4–7 | 1–2 | – | Decorative accessories, handles, bezels, etc., for pewterware; costume jewellery (rubber mould centrifugal castings) |
| – | – | 42 | Candle moulds |

It must be borne in mind that the rapid advances in plastics technology have led to a large substitution of many of the components formerly made as tin-alloy die-castings.

### Processes

The techniques used for pressure and gravity die-casting and for centrifugal rubber-mould casting (really a variation of pressure casting) are described in Chapter 7. Pressure die-casting should be carried out at the minimum pressure consistent with good reproduction of the die cavity and minimum casting porosity, since erosion of the metal die parts can be caused by turbulent flow of molten tin alloy in the small channels. The technique is mostly used for instrument cases and similar components.

Candle moulds are tin-base gravity die-castings which themselves become moulds for the die-casting of wax candles. They vary in length from a few centimetres up to 2 m and are cast from a solder alloy containing about 58 per cent tin, balance lead. The moulds are die-cast to give a smooth surface which in turn is imparted to the candle.

### 4.4 FUSIBLE ALLOYS

The term 'fusible alloy' is used to include the range of alloys which owe their primary importance to the fact that they melt at temperatures normally below the melting point of pure tin and soft solders. The alloys are based on the eutectics occurring in alloy systems formed by metals of low melting point, notably tin, bismuth, lead, indium and cadmium. Some, such as Wood's metal, were developed empirically to fill the need for a low-melting metal, but others of more recent origin were developed for specific purposes.

### Composition of fusible alloys

The compositions and melting points of selected fusible alloys and eutectic alloys containing tin are shown in Table 11.

Two groups of alloys are shown. The first is based on the ternary eutectic of tin, bismuth and lead which melts at 96 °C; Rose's alloy is one which approximates to this eutectic. The second group is based on the quaternary eutectic of tin, bismuth, lead and cadmium, melting at 70 °C; Wood's alloy is representative of this group. Of the remaining alloys, those having the highest melting points are simple binary or ternary eutectics whilst those with the lowest melting points are four- or five-element alloys containing, for example, indium. Alloys can be chosen to provide metals with fairly regularly graded melting temperatures from 20 to 227°C. While some are near-eutectic composition and have no liquidus-solidus gap, many other alloys melt over a range of temperatures and are suitable for specific applications.

### Table 11

Compositions and melting temperatures of some fusible alloys

| Type of alloy | Composition (%) | | | | | Melting range | | | |
| | | | | | | Solidus | | Liquidus | |
| | Sn | Bi | Pb | Cd | Others | °C | °F | °C | °F |
|---|---|---|---|---|---|---|---|---|---|
| Binary eutectic | 62.0 | – | 38.0 | – | – | 183 | 361 | 183 | 361 |
| Binary eutectic | 67.0 | – | – | 33.0 | – | 176 | 349 | 176 | 349 |
| Ternary eutectic | 51.2 | – | 30.6 | 18.2 | – | 145 | 293 | 145 | 293 |
| – | 48.8 | 10.2 | 41.0 | – | – | 142 | 288 | 166 | 331 |
| Binary eutectic | 43.0 | 57.0 | – | – | – | 138 | 281 | 138 | 281 |
| Commercial alloy | 58.0 | – | – | – | 42.0 In | 117 | 243 | 145 | 292 |
| Binary eutectic | 48.0 | – | – | – | 52.0 In | 117 | 243 | 117 | 243 |
| Commercial alloy | 14.5 | 48.0 | 28.5 | – | 9.0 Sb | 103 | 217 | 227 | 440 |
| Ternary eutectic | 25.9 | 53.9 | – | 20.2 | – | 103 | 217 | 103 | 217 |
| Commercial alloy | 33.0 | 34.0 | 33.0 | – | – | 96 | 205 | 143 | 289 |
| Rose's | 22.0 | 50.0 | 28.0 | – | – | 96 | 205 | 110 | 230 |
| Newton's | 18.8 | 50.0 | 31.2 | – | – | 96 | 205 | 97 | 207 |
| Ternary eutectic | 15.5 | 52.5 | 32.0 | – | – | 96 | 205 | 96 | 205 |
| Ternary eutectic | 17.0 | 57.0 | – | – | 26.0 In | 79 | 174 | 79 | 174 |
| Binary eutectic | – | 33.7 | – | – | 66.3 In | 72 | 162 | 72 | 162 |
| – | 15.4 | 38.4 | 30.8 | 15.4 | – | 70 | 158 | 97 | 207 |
| Commercial alloy | 11.3 | 42.5 | 37.7 | 8.5 | – | 70 | 158 | 90 | 194 |
| – | 13.0 | 42.0 | 35.0 | 10.0 | – | 70 | 158 | 80 | 176 |
| Wood's | 12.5 | 50.0 | 25.0 | 12.5 | – | 70 | 158 | 72 | 162 |
| Quaternary eutectic | 13.1 | 49.5 | 27.3 | 10.1 | – | 70 | 158 | 70 | 158 |
| Ternary eutectic | 12.0 | 49.0 | 18.0 | – | 21.0 In | 57 | 136 | 57 | 136 |
| Quinary eutectic | 8.3 | 44.7 | 22.6 | 5.3 | 19.1 In | 47 | 117 | 47 | 117 |
| Binary eutectic | 8.0 | – | – | – | 92.0 Ga | 20 | 68 | 20 | 68 |

Although most cast metals contract during solidification, fusible alloys containing a high proportion of bismuth do not. This very useful property arises because bismuth expands on freezing. Indeed, by controlling the bismuth content it is possible to select an alloy which will give a controlled amount of expansion on freezing or one which will retain the same volume dimensions as the mould in which it is cast.

The published literature reveals wide differences in the hardness and tensile test results obtained in various investigations with fusible alloys. Some typical results are shown in Table 12.

## Table 12

Approximate mechanical properties of some fusible alloys

| Nominal composition (%) | | | | | Tensile strength ($N/mm^2$) | Elongation (%) (slow strain) | Hardness (HB)* | Max. safe stress ($N/mm^2$) |
|---|---|---|---|---|---|---|---|---|
| Sn | Bi | Pb | Sb | Cd | | | | |
| 43.0 | 57.0 | – | – | – | 55 | 200 | 22 | 3.4 |
| 14.5 | 48.0 | 28.5 | 9.0 | – | 90 | 1 | 19 | 2.1 |
| 13.0 | 49.5 | 27.5 | – | 10.0 | 41 | 6–220 | 9–20 | 2.1 |
| 12.5 | 50.0 | 25.0 | – | 12.5 | 31 | 3 | 25 | – |
| 8.3 | 44.7 | 22.6 | – | 5.4 (+ 19 In) | 37 | 1.5 | 12 | – |

*Brinell Hardness Value

It must be emphasised that these figures provide an indication only of the properties of fusible alloys and they should not be regarded as absolute values. There is also an ageing effect so that the mechanical properties of many fusible alloys depend on the period of time which has elapsed since casting as well as on the casting and cooling conditions. The results obtained in mechanical tests are also influenced by the testing conditions; many fusible alloys have low ductility when subjected to sudden shock, but exhibit high tensile elongation under slow rates of strain.

### Applications of fusible alloys

*Fire alarms and fuses*
In fire alarm and fire-controlling systems, the restraining links that hold alarm-, water-valve- or door-operating mechanisms in position are frequently soldered with a low melting fusible alloy. Any rise in ambient temperature sufficient to melt this link results in the control system coming into operation. The ceiling-mounted water sprinklers in offices and departmental stores are a familiar application of fusible alloys, usually in this case Wood's metal. Electrical installations such as telephone exchange apparatus and electrical water-heater equipment are often protected against excessive currents by fusible-alloy heat fuses which break the circuit on reaching a certain temperature. Electric night-storage heaters also often use a fusible-alloy link as a protection against overheating. Fire detectors incorporate a joint made with one of these alloys. When the ambient temperature is high due to a fire, the joint melts and breaks, and this activates electric alarm bells.

*Temperature indicators*
A further application is for temperature indication in components where other forms of temperature measurement are not practicable, such as aircraft header

tanks, test bearings and experimental rigs. A series of rods made from fusible alloys of known melting temperature may be incorporated in the apparatus; inspection then reveals which rods melted and will thus indicate what temperatures were reached at various points.

## Seals

Fusible alloys also find use as soldering and sealing materials because the low melting temperatures will not damage the parent components. A frequently used low-melting-point solder for pewter contains 10 per cent bismuth, 49 per cent tin and 41 per cent lead, while alloys such as 50 per cent tin—50 per cent indium will wet and bond to glass. Surgical syringes have been assembled with a 44 per cent tin—28 per cent lead—27 per cent bismuth—1 per cent antimony alloy. Some types of double glazing panels incorporate fusible alloy seals.

## Work-piece and tool holding

When small objects of complex shape have to be held accurately during machining, fusible alloys offer a simple solution. The work-piece is embedded in a suitable alloy and the solid mass can then be held firmly so that the aligned component can be machined. For example, in the fabrication of compressor rotor blades for gas turbines, the blades need support whilst dovetail roots are broached. By encapsulating the blade in bismuth—tin eutectic alloy, it is held firmly and the alloy matrix serves as a jig to locate the blade accurately. After machining, the fusible alloy can be melted away and used again.

Punch-tool setting can be a slow and costly process requiring the machining of precisely dimensioned locating holes by a skilled toolmaker. Fusible alloys can be used to reduce significantly both time and cost. Apertures larger than the tool diameter are roughly cut in the punch plate, tools are positioned accurately within the holes by means of steel templates and molten fusible alloy is poured around the tools. On solidification this permanently anchors them within the punch plate during use (Fig. 21). When tools need resharpening, the alloy is easily melted and the punches removed. A typical fusible alloy for this application contains 48 per cent bismuth, 28.5 per cent lead, 14.5 per cent tin and 9 per cent antimony. This alloy does not solder the punches in, but a slight expansion on solidification holds them securely.

## Metal working

A bismuth—lead—tin—cadmium alloy which melts at about 70 °C can be used to assist in making precise bends in stainless steel or other metal tubing by providing internal support, thus preventing kinking. Since the alloy has a reasonable elongation, bends with a radius less than the tube diameter are possible. This alloy expands slightly after cooling, ensuring complete support during the bending operation; after the operation it can be melted out in hot water.

Fig. 21 — Punch tools set in their support with a fusible alloy.

A novel type of press casts its own tools in fusible alloy, thereby providing a quick and economic method of manufacturing small and medium batches of pressings. A bath of molten fusible alloy is located in the bed of the press and covered with a die plate containing an aperture cut to the dimensions of the part to be formed. A sample of the hollow sheet metal pressing which is to be produced, drilled with small holes, is lowered into the molten metal through the gap in the die plate and allowed to fill with alloy via the holes. An anchor bar attached to the ram of the press is lowered into the molten alloy filling the sample. The alloy is allowed to solidify and, when the main ram is raised, it brings up the anchor bar surrounded by the solid contents of the sample; this constitutes the punch. On removing the now empty sample pattern from the solid bath, the die cavity is left. Punch and die are finished by hand grinding and production is ready to begin, using the press as a standard hydraulic press. The sequence of operations is seen in Fig. 22.

*Casting moulds*

Prototype components can be produced by the lost wax process in which a wax model is encased in moulding sand and the wax melted out before casting. The wax models of prototypes to be made by the lost wax casting process can be

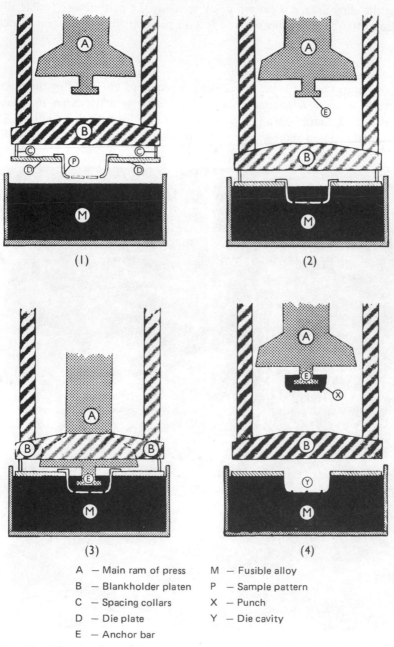

(1)                                                    (2)

(3)                                                    (4)

|   |   |   |   |
|---|---|---|---|
| A | — Main ram of press | M | — Fusible alloy |
| B | — Blankholder platen | P | — Sample pattern |
| C | — Spacing collars | X | — Punch |
| D | — Die plate | Y | — Die cavity |
| E | — Anchor bar | | |

Fig. 22 — Diagram of stages in making short-run press tools from fusible alloys by
the Dualform process.

produced by casting them in a mould of fusible alloy, which is easily made by casting the alloy around a pattern or by spraying the pattern with fusible alloy. An alloy often used for this purpose is a 60 per cent tin—40 per cent bismuth alloy.

### Fusible alloys in plastics fabrication

Enormous quantities of items in plastics are produced by moulding the plastics in suitable moulds. A comparatively recent innovation has been to spray a pattern with a bismuth—tin alloy so that a layer 1 to 7 mm thick is built up. This shell is then removed and backed up with plastics, with plaster, or with the alloy

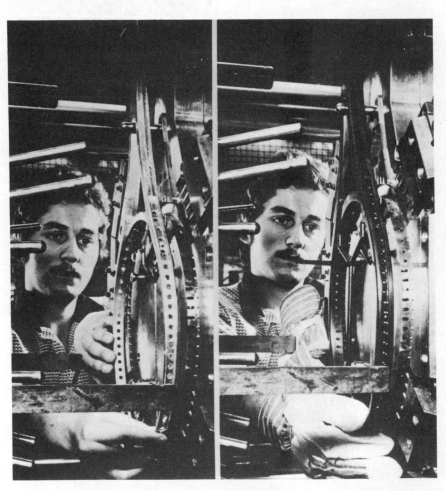

Fig. 23 — Tin—bismuth alloy core (left) used in forming tennis racquet frame from carbon-fibre reinforced thermoplastic polymer (right). (Courtesy: Dunlop Sports Co. Ltd.)

itself as the final mould. Naturally the alloy will not stand up to long runs at fast production rates, so this method is best suited to prototype and short-run production.

When resin castings and glass-reinforced plastics fabrications require a hollow cavity in them, fusible alloys are often used for the core. After moulding, these cores can be melted out, leaving the required cavity, an example being shown in Fig. 23.

## 4.5 TYPE METALS

Type face characters are small die-castings made in a lead-rich, tin-containing, alloy. The degree of perfection of the final print is of major importance, hence good casting properties of the alloy are essential. The miniature book and type illustrated in Fig. 24 exemplify the extreme detail which may be obtained.

Fig. 24 – Miniature printing block made from tin-containing type metal and the resultant printed book.

The earliest known account of casting type dates from 1540. It describes the use of the tin-rich alloys of the pewterer but it appears likely that type casting had been practised for nearly a hundred years before this and that lead and lead-rich tin–lead alloys were being used, as well as the more expensive tin-rich materials. Lead was probably too soft and, although the addition of tin improved the hardness and wear resistance, antimony was added to confer the required properties of hardness and resistance to wear and distortion under pressure.

The different kinds of type casting and printing processes require different alloy compositions, but within fairly narrow limits of the proportions of tin, antimony and lead. The criteria upon which the choice of alloy rests are hardness, resistance to wear, melting point and ability to form an accurate and sharp reproduction of the mould into which the metal is cast.

In the lead-rich corner of the ternary equilibrium diagram for tin, lead and antimony, there is a eutectic point at lead−3.8 per cent tin−11.8 per cent antimony, the melting temperature being 238 °C. This alloy has the highest fluidity of all type metals but is also relatively soft (22 HB*). The eutectic alloy has no tendency for segregation but the large specific gravity difference for the primary phases which separate in the non-eutectic alloys in this system can give non-uniform structures due to flotation of these crystals.

For the automatic, high-speed, line-composing machines of the Linotype design, the 'slugs' are cast in a near-eutectic ternary alloy in order to achieve high fluidity and minimum melting range. These machines have a keyboard somewhat similar to that of a typewriter. As each key is operated, a brass matrix having that particular character stamped on its face is released from a magazine and is assembled in its proper order. When a sufficient number of matrices have been assembled to form a line they are automatically surrounded with a mould, and a pump operates and casts a 'slug' of alloy having on its edge the line of characters forming the line of type.

The Monotype machine is quite different in principle from the Linotype machine in that each individual type character is cast separately and is assembled in order of line and column automatically. Type can be produced for storage or for hand setting if required. The functions of composing on the keyboard and of casting are separated so that they can be carried out at different times and, as the caster is entirely automatic, one man can supervise a number of casting machines.

The Monotype machine is capable of handling metals which solidify over a temperature range of about 50 °C and with liquidus temperatures not higher than 300 °C. This allows the use of harder and more wear-resistant alloys than can be used for Linotype casting, such as lead−7 to 9 per cent tin−15 to 19 per cent antimony, which have a hardness of about 27 HB and hence have good wear resistance for long production runs.

In newspaper work, printing was originally from the set type, but as circulations increased the number of printing presses required to print an edition increased and it became necessary to set replicate type for each machine. Means were sought to obtain the necessary duplicate printing plates from one master and the process of stereotyping was developed. The master type is set on either Linotype or Monotype machines and a matrix is made by pressing papier mache on to the master. The matrix is dried, removed and the type metal cast into it to

*Brinell Hardness Value

obtain the working face. Stereo plates are sometimes cast flat but more usually they are cast curved for use on rotary presses and may weigh up to 25 kg. Stereotype has a short but intensive life and therefore must be hard wearing. It is used in large pieces and therefore must be strong but, because it is cast against papier mâché, its melting point must be low. The best compromise to meet these conflicting requirements is obtained with alloys containing 15 per cent antimony and 10 per cent tin. For flat plates where maximum definition (as for half-tone blocks) is not required, the tin content may be reduced to around 5 per cent to increase the amount of primary SbSn cubes in the structure and hence improve wear resistance.

## 4.6 BRONZE

Tin bronzes (copper—tin alloys) are an important family of copper alloys and their technology is included in the monograph *Copper and its Alloys* in this series. However, they represent an important application of tin and therefore warrant attention being given to them in the present publication.

The use of tin as an alloying element with copper started with the making of artefacts and implements in the Bronze Age. Bronze still continues to be a significant tonnage use for tin and is likely to remain so in the foreseeable future. Many different copper alloys exist, some of which constitute alternative materials to tin bronze, but for many specific applications tin bronzes are still essential. Gunmetals are bronzes which contain additionally some zinc and these are widely used as casting alloys in place of binary bronzes. The use of tin bronze has remained economically viable due in part to the recycling of scrap; something like one-half of the new bronze cast in a foundry is likely to come from secondary material. Alloying with tin not only gives increased strength to copper but also imparts improved bearing performance and good corrosion resistance in certain environments.

Bronzes are utilised in three forms — as castings, as wrought material and as sintered powder components for special applications. In general the cast alloys may contain up to about 12 per cent tin but there is a demand for higher tin levels (for example 15 per cent) and for special applications, such as bells and musical cymbals, 20 to 25 per cent tin is needed to produce the required tonal qualities. Gunmetals are a cheaper substitute for certain castings, while lead to the extent of as much as 20 per cent may be present where superior bearing properties are necessary.

Wrought bronze may be in the form of rod, wire or sheet and is of lower tin content than cast bronze (maximum 9 per cent) to ensure that it can be cold worked. Phosphorus may be present in excess of that required to deoxidise the alloy during melting (see Chapter 7), to produce the harder and stronger phosphor bronzes.

A wide range of bronzes and leaded bronzes is manufactured by the powder

metallurgy route, usually by atomisation of pre-alloyed metal but sometimes by
sintering together elemental powders.

### Microstructure
The binary copper–tin equilibrium diagram is given in Fig. 25. The solid solubility

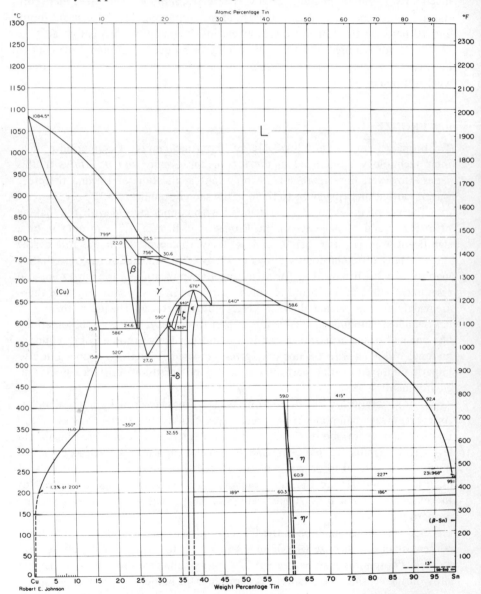

Fig. 25 – The copper–tin equilibrium diagram. (Courtesy: *Amer. Soc. Metals Handbook.*)

of tin in copper ($\alpha$ phase) is restricted to a maximum of 16.2 per cent at about 520 °C. Solid solubility falls rapidly with decrease in temperature under equilibrium conditions to about 1 per cent at 100 °C but in practice there is little difficulty in obtaining structures solely of the cored $\alpha$ solid solution (Fig. 26) even in alloys containing above 10 per cent tin.

Fig. 26 – Microstructure of single phase bronze containing 5% tin consisting of cored $\alpha$ solid solution. (Polish attack with copper ammonium chloride, etched in sodium dichromate.) (Magnification × 100.)

The production of true equilibrium structures in alloys containing tin up to the solid solubility limit is difficult and, as cast, the higher tin content alloys will normally contain some ($\alpha + \delta$) eutectoid (Fig. 27). The $\delta$ phase is an intermetallic which corresponds to the formula $Cu_{31}Sn_8$ and has a structure similar to that of $\gamma$ brass. From the practical viewpoint the $\alpha$ and $\delta$ phases are the only ones likely to be present at room temperature in most commercial bronzes.

The body-centred cubic structures $\beta$ and $\gamma$ are metastable phases which transform on slow cooling; they can be retained to some extent by quenching, more especially if the tin content of the alloy exceeds about 20 per cent. These phases are ductile which is of practical importance in connection with the hot working of bronzes. The $\zeta$ and $\eta$ phases have hexagonal structures, corresponding to $Cu_{20}Sn_6$ and $Cu_6Sn_5$ respectively, the latter being regarded as a superlattice based on a nickel arsenide structure. These phases are unlikely to be found in commercially produced bronzes although $\eta$ is formed by a diffusion reaction of copper with a solid or liquid tin coating. The $\epsilon$ phase has an orthorhombic structure and corresponds to the composition $Cu_3Sn$. When saturated $\alpha$ or ($\alpha + \delta$) structures are annealed for a long time below 350 °C, $\epsilon$ very slowly precipitates.

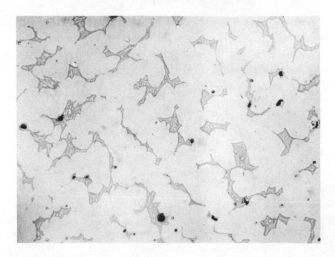

Fig. 27 – Microstructure of bronze containing 14% tin showing presence of (α + δ)eutectoid in α solid solution. (Etched in sodium dichromate.) (Magnification ×150.)

However it is never likely to form in commercial bronzes unless the service conditions involve prolonged heating in the range 200 to 350 °C. Traces of ε are also seen in hot-tinned copper between the coating layer and the basis metal.

Changes in the solid solubility of tin in copper below 500 °C proceed so sluggishly that as far as the normal production, heat treatment and fabrication of bronze are concerned it is permissible to modify the equilibrium diagram to show the δ phase existing down to room temperature and the α phase boundary as a vertical line at 16 per cent tin extending from just below 600 °C to room temperature. Annealing at or above 500 °C followed by slow cooling retains 13 to 16 per cent of tin in solid solution, the exact amount retained being dependent on the time and temperature.

The long solidification range of α alloys leads to considerable coring and, even if no δ is formed, heat treatment for several hours at about 600 °C may be necessary to homogenise the alloy, especially if it is to be subjected to any fabrication process. Bronzes, however, are often used in the as-cast condition and a non-equilibrium structure in which δ is present is frequently acceptable. When the tin content is high (for example 20 per cent) the as-cast structure may show massive δ phase (Fig. 28) and in this condition the alloy is brittle. Prolonged soaking at 700 to 750 °C followed by quenching will lead to softening by retaining an α→β structure (Fig. 29) which allows working of the alloy. Subsequent heat treatment causes partial decomposition of the β phase resulting in precipitation hardening and use is made of this effect in the making of some musical cymbals.

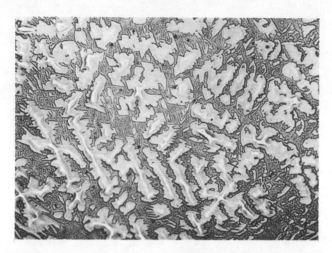

Fig. 28 – Microstructure of cast bronze containing 20% tin showing presence of massive δ phase and α solid solution. (Etched in sodium dichromate.) (Magnification × 300.)

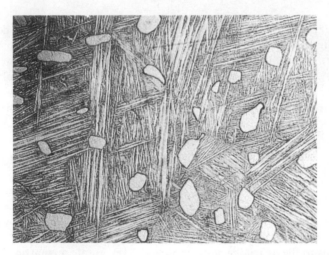

Fig. 29 – Microstructure of 20% tin bronze water-quenched from 700 °C to produce α solid solution containing partially decomposed martensitic β phase. (Etched in ferric chloride.) (Magnification × 150.)

Phosphorus is widely used as a deoxidant in bronzes but higher phosphorus contents (*ca* 1.5 per cent) are sometimes deliberately introduced to form phosphor bronzes containing the hard compound $Cu_3P$ in the structure in order to improve wear resistance (Fig. 30). The association of $Cu_3P$ with the $(\alpha + \delta)$ eutectoid should be noted.

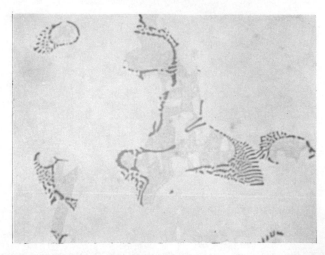

Fig. 30 — Microstructure of phosphor bronze (10% Sn–0.5% P) showing association of $Cu_3P$ (dark) with the $(\alpha + \delta)$ eutectoid. (Etched in sodium dichromate.) (Magnification × 1500.)

When zinc is added to bronze to form gunmetals it enters into solid solution in the $\alpha$ phase. In rapidly cooled alloys the $\alpha$ range extends from approximately 5 per cent tin, no zinc to 30 per cent zinc, no tin. With the zinc contents customarily employed in gunmetals, completely $\alpha$ structures can be obtained by appropriate homogenisation.

Lead is often added to bronze and gunmetal to improve machinability and particularly bearing properties. Lead is virtually insoluble in solid copper and only partly miscible in liquid copper so that the lead is present as a separate phase. The uniformity of its dispersion will depend on the casting conditions and the habit of lead in a typical gunmetal is shown in Fig. 31.

### Mechanical properties

The tensile strength of chill cast bronzes increases linearly with tin content up to 5 per cent and then more gradually, levelling out at about 10 per cent tin. This corresponds to the increasing amount of $(\alpha + \delta)$ eutectoid present and is followed by a fall in strength, with the formation of continuous intercrystalline films of $\delta$. The yield point, however, continues to increase, reaching a maximum at about 20 per cent tin, where it approximates to the tensile strength. The ductility, as indicated by the tensile elongation, reaches a maximum towards the end of the $\alpha$ range and then falls to virtually zero with 25 per cent tin content. Hardness increases linearly with tin content through the $\alpha$ range and thereafter more rapidly to 300 HB at 25 per cent tin. Phosphorus up to 0.8 per cent appears to have no effect on the elongation of these alloys, but, like tin, increases hardness, being generally more effective than tin in this respect.

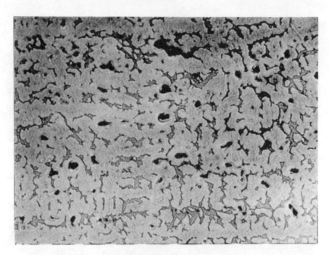

Fig. 31 – Microstructure of leaded bronze (10% tin + 10% lead) showing uniform distribution of the insoluble lead particles. (Etched in sodium dichromate.) (Magnification ×200.)

The mechanical properties show a fall in tensile strength and hardness as the lead concentration of bronze increases. Nickel is added to bronze to improve mechanical properties of sand castings and to decrease cost. It also refines grain size and reduces the freezing range, thus providing more latitude in casting temperature.

An α-bronze may be worked into sheet and its resultant mechanical properties depend on the final cold-rolling reduction. Very severe cold working of 5 per cent tin bronze (for example over 95 per cent cold reduction) has been shown to give a spinoidal decomposition of the structure during annealing and very favourable mechanical properties.

The overall corrosion resistance of bronzes and gunmetals is often a major reason for their use as valves, pump bodies, condenser tubing, gauze filter meshes for paper making, marine fittings, etc.

Copper–9 per cent nickel–2 per cent tin is a recently developed electrical spring material which has good oxidation resistance and hence is easy to solder. The higher nickel content alloys (for example copper–6 to 20 per cent tin–6 to 20 per cent nickel) have been developed for castings which can be heat treated to improve mechanical properties by precipitation, but these are not in wide commercial use.

A new alloy has been developed containing copper–5 per cent tin–1 per cent magnesium which is solution treated, quenched and then heat treated at 375 °C to provide mechanical properties and electrical conductivity similar to a beryllium–copper alloy. This results from the precipitation of the phase $Cu_4SnMg$ on crystallographic planes or on deformation bands in cold worked material

(Fig. 32) to provide a material combining good electrical conductivity with strength and hardness. It is not yet available commercially.

Fig. 32 – Microstructure of worked and fully heat treated copper, 5% tin, 1% magnesium alloy showing precipitation of $Cu_4SnMg$ on deformation planes causing hardening. (Polish attack with ammonium persulphate and hydrogen peroxide.) (Magnification × 500.)

Other recent developments include an alloy with 5 per cent tin and 17 per cent aluminium which has high staining resistance but is easier to hot fabricate than to cold work.

## Applications of Bronze
The main reasons for using tin bronzes and gunmetals are their combination of a number of desirable properties:

(1) High mechanical strength and hardness.
(2) Reasonable electrical conductivity.
(3) Good casting and/or working properties.
(4) Resistance to corrosion.
(5) Excellent bearing characteristics.
(6) Ease of soldering.

A combination of factors (3) and (4) is made use of in cast valves, etc., while (5) applies particularly to the alloys containing lead. Electrical springs and wires require factors (1), (2) and (3).

The status of bronze billets for subsequent machining into components was greatly enhanced by the virtually complete change from the use of chill cast billets to continuously cast rod and strip with effectively no scrap losses. The mechanical properties of continuously cast bronze are noticeably superior because of the rapid chill, controlled casting conditions and lack of porosity. The development of this process allowed a new class of materials to be introduced into BS 1400 in 1961.

*Bronze castings*
The melting and casting procedures for these materials are described in Chapter 7. The tin-containing copper alloys for castings included in BS 1400: 1973, are shown with their principal properties and applications in Table 13.

**Table 13**

Some tin-containing copper casting alloys in BS 1400:1973

| Sn | P | Zn | Pb | BS.1400:1973 grade | Minimum tensile strength $(N/mm^2)$ (test bars[1]) | Principal applications (see key below) |
|---|---|---|---|---|---|---|
| 12.0 | 0.15 min. | — | — | PB 2 | 220 | B(H), G(H) |
| 10.0 | 0.4 min. | — | — | PB 4 | 190 | B(H), G(H), C |
| 10.0 | 0.15 max. | — | — | CT 1 | 230 | C |
| 10.0 | — | 1 max. | 10.0 | LB 2 | 190 | B(M), C |
| 9.0 | — | 1 max. | 15.0 | LB 1[2] | 170 | B(L) |
| 7.0 | — | 2.5 | 3.0 | LG 4 | 250 | B(L), P |
| 5.0 | — | 1 max. | 21.0 | LB 5[2,3] | 160 | B(L) |
| 5.0 | — | 5 | 5.0 | LG 2 | 200 | B(L), G(L), P |
| 7.5 | 0.3 min. | 2 max. | 3.5 | LPB 1 | 190 | B(M) |

*Key to application areas*
B(L)  Low loaded bearings and for unhardened steel shafts.
B(M)  Moderately loaded bearings perhaps with less than perfect lubrication.
B(H)  Highly loaded bearings with good lubrication perhaps with pounding loads.
C     Corrosion-resistant cast valves and pumps for sea and boiler feed water and mineral acids.
G     Gears and wormwheels with high (H) or low (L) loading.
P     Pressure-tight castings.

*Notes*
(1)  From sand cast material; for details on properties of materials cast by other methods, see the full British Standard specification.
(2)  Specially for poor lubrication conditions.
(3)  Also contains 5.5% Ni; other bronzes allow mostly 1–2% Ni.

Tin bronzes are used in the as-cast condition usually because they combine corrosion resistance, wear resistance and strength, properties which are particularly important for many engineering uses, as illustrated in Fig. 33. In general the more severe the expected mechanical service conditions the higher the tin content used, an example being a 15 per cent tin content alloy (perhaps with 0.75 per cent phosphorus) for heavily loaded bearing bushes. Where some shock resistance is necessary, the tin content should be lower to provide an alloy with greater ductility (for example, 13 per cent tin, 0.5 per cent phosphorus maximum). Tin contents of 10 to 12 per cent are commonly used for centrifugal casting of gear wheel blanks, while hard 20 per cent tin bronzes have been used for bridge bearings. The single-phase cast alloys with around 8 to 10 per cent tin content are ideal for valves for certain chemical plant or for similar environments where corrosion resistance is important.

Fig. 33 – A selection of cast tin-bronze engineering components.

The bearings properties of bronzes improve fairly rapidly as the lead content is increased from 5 to about 15 per cent and thereafter more slowly as it approaches 25 per cent. There is little difference between the leaded gunmetals and leaded

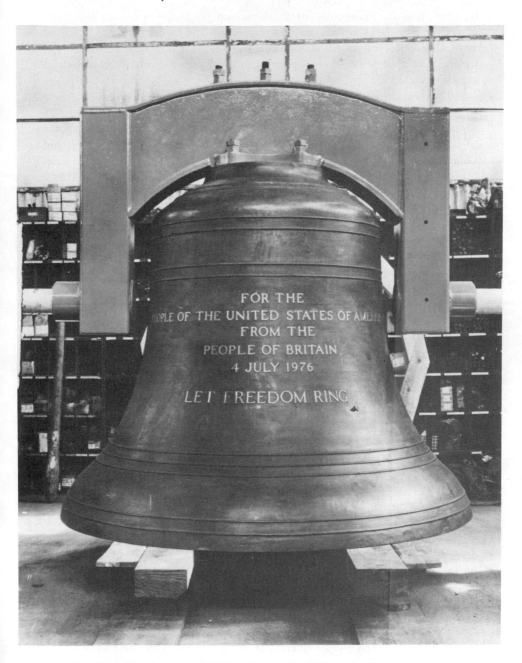

Fig. 34 – Cast bronze bell. This replica of the "Liberty Bell" was cast to commemorate the bicentennial anniversary of the inauguration of the United States of America.

bronzes from the standpoint of their bearing properties, although the latter alloys are considered to be somewhat superior. The benefits derived from the addition of lead are gained at the expense of reduction in wear resistance and hardness so that leaded bronzes and gunmetals are not used for high load conditions.

Although the addition of zinc to form gunmetals lowers corrosion resistance to some extent, ease of casting and soundness of casting are generally improved, so that the alloys of the types BS 1400: LG2 and LG4 are widely used in place of binary tin bronze for applications such as valves and fittings for water and steam pipelines and for bearing shells.

Bronzes are the widely accepted materials for statuary and artistic castings, which are normally cast in sand moulds. Large units may be cast in sections and subsequently joined together. The tin content of the binary bronzes used for these purposes is usually 8 to 12 per cent, perhaps with a small percentage of zinc, and the surface develops a pleasing, dark, protective patina during exposure to the atmosphere.

One very old and specialised use of high-tin bronzes is for the manufacture of bells, for which binary alloys containing about 24 per cent tin are almost invariably used and this composition has not changed for hundreds of years. Bells are cast in loam moulds and the casting technique is sufficiently good that no finishing work has to be done on the outer surface (Fig. 34). Tuning is effected by paring away metal from the inner surface. A similar type of alloy, usually of slightly lower tin content, is used for making cymbals.

*Wrought bronze*
The compositions and properties of the principal tin-containing copper alloys in wrought form included in the relevant British Standards are given in Table 14.

**Table 14**

Tin-containing wrought copper alloys

| Nominal % Sn | BS and grade† | Typical tensile strength (N/mm²) |
|---|---|---|
| 9 | BS 2871:1957 (PB104) | 460 (drawn tube) |
| 7 | BS 2870:1980 (PB103) | 500 (half-hard sheet) |
| 5 | BS 2870:1980 (PB102) | 490 (half-hard sheet) |
| 3 | BS 2870:1980 (PB101) | 420 (half-hard sheet) |

† See also BS 2873, BS 2874, BS 2875 covering the same alloys in the forms of rod, wire and plate respectively.

Fabrication of bronzes is generally only possible for the lower tin contents in which the brittle ($\alpha + \delta$) eutectoid is virtually absent. Thus the bulk of wrought bronze falls in the 5 to 8 per cent tin range and these alloys may be worked into sheet, strip, tube and wire, the latter being made by drawing down from continuously cast rod or hot forged stock. Hot rolling of bronze into sheet is not usually undertaken as a viable commercial process except for tin contents in the region of 3 to 4 per cent. Coinage bronze contains less than 1 per cent tin in the United Kingdom.

More recently developed wrought alloys are the copper–2 per cent tin–9 per cent nickel electrical spring material (CA 725), the copper–8 per cent tin–1 per cent aluminium alloy suitable for use in seawater desalination plants, the copper–5 per cent tin–7 per cent aluminium alloy proposed as having good tarnish resistance for architectural applications, and the heat-treatable copper–5 per cent tin–1 per cent magnesium alloy.

*Sintered bronze*

Repetition parts such as small bearing bushes are not only made by machining from cast bronze rod or from wrought strip but are also produced by the powder metallurgy route. Mixed elemental copper and tin powders are pressed into a die of the required shape and size and the 'green' compact is then sintered in a reducing atmosphere at around 500 °C to produce a porous but relatively strong bearing bush or similar component. This may be subsequently impregnated with oil to produce a bearing requiring little or no further lubrication during service. A 10 per cent tin content would be typical and lead or graphite may also be incorporated into these materials to provide improved lubricity. Sintered filters and spark arresters for gas lines are usually made by compacting and sintering pre-alloyed bronze powder, produced as spherical particles of the required diameter for a given porosity.

One process for the manufacture of steel-backed bronze strip for forming into shell bearings consists of sintering and rolling bronze powder on a steel strip by a continuous process.

## 4.7 TIN IN SINTERED IRON

An important and expanding industry is the mass production of small engineering components by the powder metallurgy route. Iron powder is pressed in a die to give a part of the required shape, and this is then sintered at an elevated temperature to give the final component. To enable the sintering operation to be carried out at a temperature significantly below the melting point of iron, the powder is frequently alloyed with, for example, copper. Research work has demonstrated, however, that tin additions can be used with advantage to lower the required sintering temperature. Detailed studies with various tin plus copper additions demonstrated that properties comparable with those obtained by sintering iron powder plus 10 per cent copper at 1150 °C could be achieved at only 950 °C with an alternative addition of 2 per cent tin plus 3 per cent copper. This lower sintering temperature enables a cheaper and more reliable type of furnace to be employed, reduces the energy required and also reduces the degree of distortion during sintering.

An alternative but related use of tin additions to sintered iron parts is for the control of dimensional tolerances in iron compacts containing, say, 10 per cent copper without adding sufficient tin to gain the advantage of a lower sintering temperature. During sintering, iron parts containing copper tend to grow as the copper dissolves in the iron, whereas the addition of tin reduces the copper solubility and thereby reduces the dimensional change. The addition of 1 per cent tin can therefore be beneficial when the furnace equipment is being operated at the higher temperature because of the range of different types of components being manufactured.

There are already manufacturers in many different countries adding tin to sintered iron parts; Fig. 35 shows a selection of these components. It is interesting to note that this manufacturer, using the iron—tin—copper mix for the manufacture of pistons and connecting rods used in domestic refrigerators, also

Fig. 35 — Compressor piston components for domestic refrigerators made from
sintered iron with the addition of tin as a sintering aid.

claims to obtain improved wear resistance. This no doubt arises from the presence of the hard copper—tin phase in the final part.

## 4.8  TIN ALLOYED CAST IRON

Cast iron is the most widely used material in the engineering industry since it combines low cost, good castability and corrosion resistance with tensile strength sufficient for general engineering use.

Experimental work has proved that the addition of about 0.1 per cent tin to flake graphite or nodular cast iron is extremely potent in promoting the formation of a fully pearlitic structure in the matrix. This improves the wear resistance, reduces the dependence of hardness on section size and enhances machinability, thus reducing machine tool wear. The practice of making tin additions to cast iron is now widely accepted in industry throughout the world as a simple process which can be employed for the production of high quality castings in both mechanised and non-mechanised iron foundries.

The tin is added as pellets or stick to molten iron flowing from the cupola launder, or to the ladle of iron as it is filled (Fig. 36). Since tin melts at 232 °C

Fig. 36 — Adding sticks of tin to cast iron as the ladle is filled to ensure a pearlitic as-cast structure in the castings.

Fig. 37 – Microstructure of grey cast-iron, thin-walled piston made in a shell mould (above) showing surface layers of massive ferrite and (below) the same casting made with the addition of 0.1% tin to the iron to produce a completely pearlitic, wear-resistant structure. (Etched in 2% Nital.) (Magnification ×85.)

it is instantly melted and dissolved in the molten cast iron in which it has a high liquid and solid solubility. On the other hand, tin has a high boiling point, so that virtually no tin is lost by vaporisation when it is added to the iron, even at these high temperatures of around 1400 °C.

Unlike other pearlite promoters such as chromium, there is no tendency to form hard carbide particles or to increase the depth of chill, so that machining properties are unimpaired. The normal mode of use, therefore, is to determine the iron composition which provides a fully pearlitic structure in the thinnest section to be machined and to add tin to prevent massive ferrite forming in adjoining thick sections.

A special application is for castings made in shell moulds where the particular nature of the mould induces fine graphite and associated surface layers of ferrite. The latter are virtually eliminated by prior alloying of the iron with tin (Fig. 37).

Another property of tin when added to cast iron is that it stabilises the pearlitic structure of the iron at high temperatures and counteracts the natural tendency of iron to soften in prolonged high temperature applications. This softening is due to decomposition to a ferritic structure and there is an associated dimensional growth of the castings which is often unacceptable; this is reduced by the presence of tin (Fig. 38). The pearlite stabilisation is also of importance

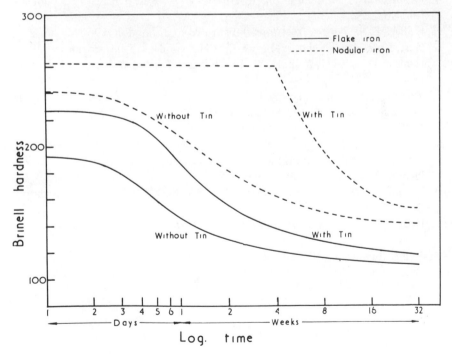

Fig. 38 — Effect of 0.1% tin in flake graphite and nodular graphite cast iron on the resistance to softening by pearlite decomposition at 500 °C.

in maintaining a uniform hardness irrespective of the time that castings remain in the mould.

A feature of tin-alloyed cast iron is that the tin is not lost on remelting, so that in practice foundry scrap being recycled provides part of the necessary tin and the balance is added as tin metal immediately prior to casting. The actual percentage added might therefore only be, say, 0.06 per cent, but of course the tonnage of cast iron produced is very high. It may also serve as a convenient way of re-using tin-bearing steel scrap.

Recent examples of the use of tin in cast iron include motor car cylinder blocks, crankshafts, axles for commercial vehicles, brake drums, transmission components, hydraulic lifting equipment and many other industrial applications.

## FURTHER READING

1. *The Properties of Tin Alloys*, International Tin Research Institute Publication No. 155.
2. *Working with Pewter*, International Tin Research Institute Publication No. 566.
3. Charron, S., *Modern Pewter*, Van Nostrand Reinhold, New York, 1973.
4. *Fusible Alloys containing Tin*, International Tin Research Institute Publication No. 175.
5. Hedges, E. S., *Tin and its Alloys*, Edward Arnold, London, 1960.
6. West, E. G., *Copper and its Alloys*, Ellis Horwood Ltd (this series), 1982.
7. Hanson, D. and Pell-Walpole, W. T., *Chill-cast Tin Bronzes*, Edward Arnold, London, 1951.
8. Plummer, D. G. and MacKay, C. A., Optimization of properties of a precipitation hardening tin—magnesium bronze, *Metals Technology*, Oct., p. 457, 1977.
9. Long, J. B. and Robins, D. A., Improving the sintering performance of iron powder by tin additions, *Modern Developments in Powder Metallurgy*, Plenum Press, Vol. 4, p. 303, 1971.
10. *Tin-alloyed Iron Castings*, International Tin Research Institute Publication No. 545.
11. Hedges, E. S., *Tin in Social and Economic History*, Edward Arnold, London, 1964.
12. *Metals Handbook*, 8th edn., Vol. 3 Machining, American Society for Metals, Ohio, 1967.

# 5

# Solder Alloys

---

Soft solders constitute the second largest use of primary tin, accounting for about 23 per cent of the total. In the last 20 years or so there has been a changing pattern of technology in developed countries. The established, traditional uses of solder, as in plumbing, have decreased and the soldered side-seams of cans may now contain one-twentieth of the tin content used two decades ago, but the sophisticated electronics industry, embracing sectors such as computers, communications and television, has given rise to a whole new field of soldering technology where high tin solders are universally used. The soldered joints in electronic assemblies have become smaller in size as this new technology has rapidly advanced, but this reduced quantity of solder in each joint is compensated for by the many more pieces of electronic equipment that are made, such as calculators and electronic controls in industrial processes. Modern techniques like mechanised soldering of printed circuits allow manufacturers to make several hundred connections simultaneously, amounting to millions of soldered joints of high reliability each week. Although it is important to describe in detail the metallurgy and properties of soft solder alloys, it is relevant to include in this chapter a discussion of soldering processes and of the requirements for making good soldered joints.

## 5.1 METALLURGY OF SOLDER ALLOYS

The vast majority of soft solders used are based on the binary system of tin and lead. Other alloys are based on tin and antimony, tin and silver, and the ternary lead−tin−silver system. There are a few other special solders which are also mentioned below.

### Tin−lead
The binary mixture of tin and lead constitutes a simple, classic, eutectic system. The eutectic composition is 61.9 per cent tin; the eutectic temperature is 183 °C (Fig. 39).

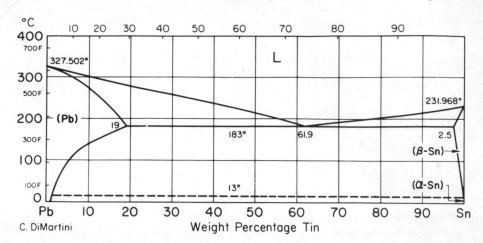

Fig. 39 – The tin–lead equilibrium diagram.
(Courtesy: *Amer. Soc. Metals Handbook*.)

The microstructures of cast alloys consist of primary dendrites of tin-rich or lead-rich solid solutions in a lamellar or globular (depending on the cooling rate) eutectic matrix consisting of both solid solutions (Fig. 40).

The temperature difference between the liquidus and the solidus, i.e. the liquidus–solidus gap, or 'pasty range', is a maximum at the lead-rich end, about 100 °C at 19 per cent tin, 81 per cent lead, whilst at the tin-rich end (at 97.5 per cent tin, 2.5 per cent lead) the range is about 40 °C. Thus all tin–lead solders remote from the eutectic composition will have a pasty range, and in commercially used alloys this can be as wide as 90 °C. It is therefore incorrect to talk of a 'melting point' for tin–lead alloys and usually from the practical joint-making aspect it is the liquidus temperature which should be considered. In practice, however, the term 'melting point' is often used when referring to the liquidus temperature. The majority of the alloys commence melting at the eutectic temperature of 183 °C and this dictates their effective service capability if the criterion is strength at elevated temperatures.

Because of the change in solid solubility of tin in lead with temperature, some softening and a decrease in strength of chill-cast alloys may be observed due to precipitation of a tin-rich phase from the primary lead dendrites. This effect is even more pronounced in the tin–lead solders containing a few per cent of antimony, since precipitation of the intermetallic compound SbSn occurs.

*Tin–antimony*
The metallurgical equilibrium diagram for the binary tin–antimony alloy system is seen in Fig. 41. From the practical point of view, the relevant portion extends

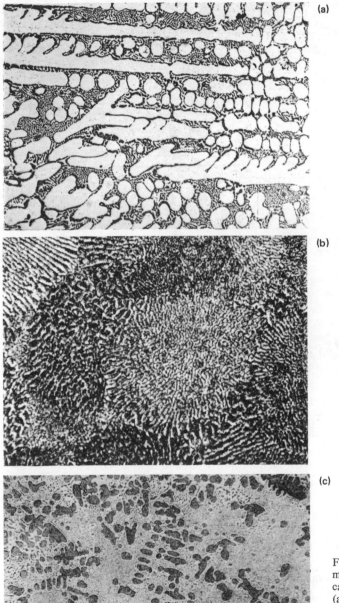

(a)

(b)

(c)

Fig. 40 – Typical microstructures of cast tin–lead alloys: (a) primary tin dendrites in 70% tin alloy; (b) eutectic alloy, 62% tin; (c) lead-rich dendrites in a eutectic matrix in 50% tin alloy. (Etched in 2% Nital.) (Magnification × 100.)

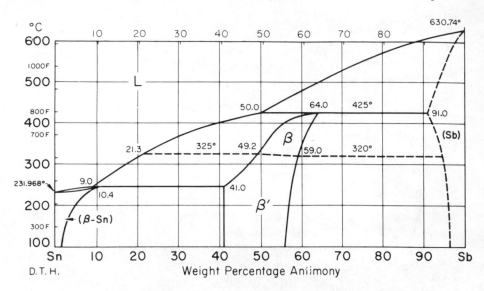

Fig. 41 – The tin–antimony equilibrium diagram. (Courtesy: *Amer. Soc. Metals Handbook.*)

only up to 5 per cent antimony by weight, indicating that these alloys are single-phase solid solutions. This practical restriction arises because a higher concentration will result in the formation of primary intermetallic SbSn cuboids during solidification to an extent dependent on cooling rate. Because of the inherently brittle nature of intermetallics, their presence is likely to be detrimental to joint strength. The high solid solubility of antimony in tin gives valuable hardening and strengthening. The change in solubility, however, to below 1 per cent at room temperature allows precipitation of SbSn, so that eventual softening of the alloys may occur during service.

### Tin–lead–antimony

In certain grades of tin–lead solder alloy, a small percentage of antimony is present, the limit being 7 per cent of the tin content to avoid formation of SbSn cuboids. The tin–lead–antimony ternary equilibrium diagram indicates that, for the most commonly used 'antimonial' tin–lead solders, with tin contents in the range 30 to 50 per cent tin, the liquidus temperature is slightly below that of the corresponding binary tin–lead alloy. At the same time the presence of antimony in such alloys raises the solidus temperature by only 2 °C, i.e. to 185 °C.

### Tin–silver

These solders have basically greater strength than tin–lead alloys. The tin–silver

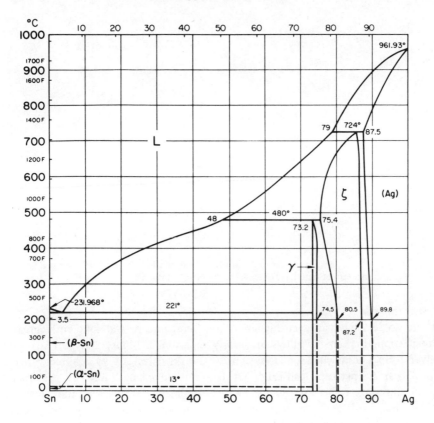

Fig. 42 – The tin–silver equilibrium diagram. (Courtesy: *Amer. Soc. Metals Handbook.*)

equilibrium diagram shown in Fig. 42 indicates the presence in this binary system of a eutectic at a composition of 96.5 per cent tin, 3.5 per cent silver consisting of the intermetallic $Ag_3Sn$ and a tin-rich phase. The eutectic temperature is 221 °C and, as shown in Fig. 42, there is a steep rise in liquidus temperature which effectively limits the silver content of alloys for practical use as solders to about 5 per cent. The structure of this alloy consists of dendrites of primary $Ag_3Sn$ in a matrix of $Sn–Ag_3Sn$ eutectic (Fig. 43).

Binary tin–silver solders that are commercially used contain between 2 and 5 per cent silver, those corresponding to or exceeding the eutectic composition having the greatest strength due to the dispersed particles of intermetallic compound. The presence of silver in tin has a considerable hardening effect, even below the eutectic composition and despite the low solid solubility (below 0.1 per cent) of silver in tin at room temperature.

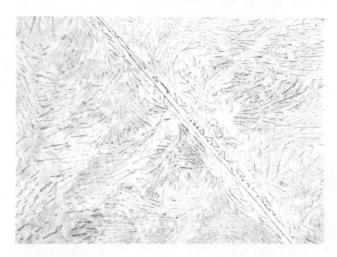

Fig. 43 – Microstructure of cast 95% tin, 5% silver alloy showing primary Ag₃Sn in a eutectic matrix. (Etched in 2% Nital.) (Magnification ×100.)

*Lead—tin—silver*
These solders have an even higher solidus temperature and inherently greater creep strength than the alloys previously mentioned. Binary alloys of lead and silver form a somewhat similar system to the tin—silver alloys, a eutectic being formed at 2.5 per cent lead (304 °C). It is not surprising, therefore, that lead—silver alloys are sometimes used as high-strength solders. However, they have the disadvantage of poor wetting properties and so, to improve wetting, small amounts of tin are added, although in insufficient quantities to form excessive amounts of the intermetallic $Ag_3Sn$. The microstructures of the commonly used solders, based on lead plus 1.5 per cent silver with either 1 or 5 per cent tin, consist of a ternary lead—tin—silver eutectic with primary lead-rich dendrites (Fig. 44).

*Tin—zinc*
Tin—zinc alloys which are specifically for soldering aluminium have structures related to the binary equilibrium diagram seen in Fig. 45. This represents a simple eutectic system with the eutectic point at 8.9 per cent zinc (balance tin) and a melting temperature of 198 °C. Alloys containing more than about 9 per cent zinc, such as the 20 and 30 per cent zinc alloys used industrially, will, upon cooling, exhibit primary dendrites of hard zinc-rich phase in a tin—zinc eutectic matrix and, because of the high hardness of zinc relative to tin, hardness will increase rapidly with zinc content above the eutectic composition.

*Low melting point solder alloys*
For heat-sensitive work, the most commonly used alloy that features in several

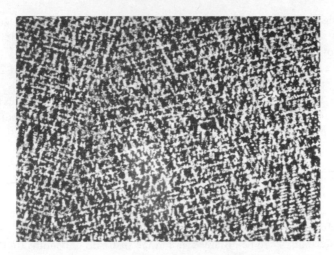

Fig. 44 – Microstructure of cast lead, 5% tin, 1.5% silver alloy showing primary lead dendrites in a eutectic matrix. (Etched in 2% Nital.) (Magnification × 100.)

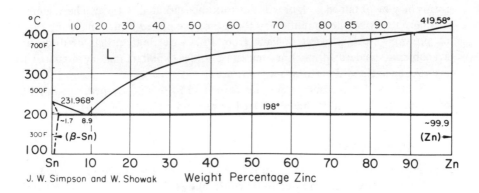

Fig. 45 – The tin–zinc equilibrium diagram. (Courtesy: *Amer. Soc. Metals Handbook.*)

specifications is the ternary eutectic of composition 50 per cent tin, 32 per cent lead and 18 per cent cadmium, melting at 145 °C.

The binary eutectic of tin and indium (48 per cent tin), melting at 117 °C, is also used in certain special applications, especially for soldering glazed surfaces.

*Other solders*

Tin–lead solder of composition near to the eutectic and saturated with copper additions (about 1.5 per cent copper) is used to reduce the attack of solder on unclad copper soldering iron bits.

Silver-loaded tin—lead eutectic alloy (62 per cent tin—36 per cent lead—2 per cent silver) is used when soldering silver-coated surfaces in order to reduce the rate of dissolution of the silver layer.

A gold—tin eutectic containing 20 per cent tin, which melts at 280 °C, has a special application in semiconductor soldering.

Pure tin is sometimes used as a solder for side seams of tinplate cans for certain products where the regulations stipulate that the hazard of lead contamination must be minimised.

Other alloys mentioned in commercial literature or in specifications are based on the tin—cadmium, cadmium—silver or tin—cadmium—silver systems. The majority of these have minor usages and are specifically for obtaining higher strength soldered joints.

### Forms of solder

Most solder alloys can be obtained in a range of forms depending on the application. For dip-soldering, ingots are normally supplied, whilst for manual sheet-metal work many different types of sticks and bars are available.

For automatic soldering, wire or ribbon solder is commonly used. These forms may be obtained with or without flux incorporated in them. The assembly of many items can be simplified by the use of pre-placed solder and many types of preforms, such as washers made from foil, wire rings, pellets or discs, are available as standard forms. For long runs, solder suppliers may be prepared to produce specialised preform shapes to order.

Solder is also available in powder form and in powder suspensions in paste or liquid fluxes, the latter being referred to as 'solder paint' (Fig. 46).


## 5.2 APPLICATIONS

The most commonly used solder alloys are binary tin—lead alloys (Table 15).

For most electrical and electronic work, alloys close to the tin—lead eutectic composition are used to obtain the benefit of the low melting point combined with maximum wetting power. For less exacting work, lower tin contents may be utilised.

Above about 100 °C, tin—lead alloys decrease noticeably in strength because they are only 80 °C or so below their solidus temperature (i.e. at 82 per cent of the fusion point on the absolute temperature scale). Alternative materials must therefore be selected for service at elevated temperatures. These may be either the 95 per cent tin—5 per cent antimony alloy, or the eutectic tin—silver alloy (3.5 per cent silver). Alternatively, the lead-based alloys with 1.5 per cent silver and 1 or 5 per cent tin are used, melting at close to 300 °C, but these have rather poor wetting properties because of the low tin content. These alloys are useful, for example, in the soldering of commutators for electric traction motors where high initial starting currents lead to rapid rises in temperature.

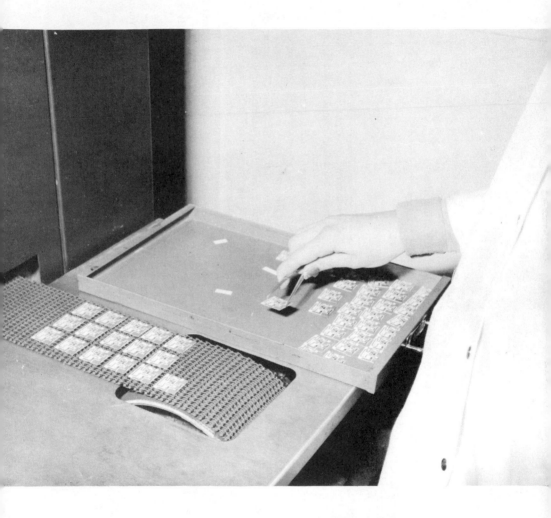

Fig. 46 – Microcircuits based on alumina substrates, being passed through an oven to reflow solder paste applied by screen-printing to specific metallised areas to provide a solderable finish where required.

**Table 15**

Some commercially used soft-soldering alloys

| Sn | Pb | Sb | Ag | Other | Nominal melting point or range (°C) | Grade in British Standard | Typical uses |
|---|---|---|---|---|---|---|---|
| 100 | – | – | – | – | 232 | 3252 Grade T2 | Can side seams. Creep resistant joints. Non-toxic. |
| 64 | Rem | 0.6/0.2 | – | – | 183–185 | 219 Grades A, AP } | High grade electrical, electronic and instrument work. Can side seams. |
| 60 | Rem | 0.5/0.2 | – | – | 183–188 | 219 Grades K, KP } | |
| 50 | Rem | 0.5 | – | – | 183–212 | 219 Grade F | Sheet metal work and light engineering. |
| 40 | Rem | 0.4 | – | – | 183–234 | 219 Grade G | General engineering and capillary fittings. Can side seams. |
| 30 | Rem | 0.3 | – | – | 183–255 | 219 Grade J | Plumber's solder, cable jointing. Motor car radiators. |
| 20 | Rem | 0.2 | – | – | 183–276 | 219 Grade V | Motor car radiators. Electric lamp bases. |
| 10 | Rem | 0.5 | – | – | 267–301 | – | Cryogenic equipment. Thin film circuits. |
| 2 | Rem | – | – | – | 320–325 | – | Can side seams. |
| 50 | Rem | 2.5–3.0 | – | – | 185–204 | 219 Grade B } | Slightly cheaper and stronger versions of the non-antimonial BS 219 Grades F, G and J, but only suitable for use on zinc-free substrates. |
| 40 | Rem | 2.0–2.4 | – | – | 185–227 | 219 Grade C } | |
| 30 | Rem | 1.5–1.8 | – | – | 185–248 | 219 Grade D } | |
| 62 | Rem | 0.2 | 2 | – | 178–184 | 219 Grade 62S | Reduced rate of attack on silver substrates. Higher creep strength than SnPb40. |
| 95 | – | 5 | – | – | 236–243 | 219 Grade 95A | { Elevated temperature applications – resistance to creep. |
| 96.5 | – | – | 3.5 | – | 221 | 219 Grade 96S | { Elevated temperature applications – resistance to creep. Reduced rate of attack on silver substrates. Non-toxic. |
| 1 | 97.5 | – | 1.5 | – | 309–310 | 219 (1959) Grade 1S } | Equipment operating at either elevated or cryogenic temperatures. |
| 5 | 93.5 | – | 1.5 | – | 296–301 | 219 Grade 5S } | |
| 30 | Rem | – | – | Cd 18 | 145 | 219 Grade T | Avoidance of heat damage to insulation and for adjacent joints made with higher m.p. solder. |
| 80 | – | – | – | Zn 20 | 200–270 | – | Soldering aluminium. |
| 48 | – | – | – | In 52 | 117 | – | Sealing glass and glazed ceramics. |

Solders generally become stronger at lower temperatures, but for very low temperature cryogenic work lead-rich solders containing at the most 20 per cent tin are necessary, because materials containing a high proportion of tin then have low ductility and poor resistance to impact. The effect of temperature on the tensile properties of different solder alloys is shown in Fig. 47.

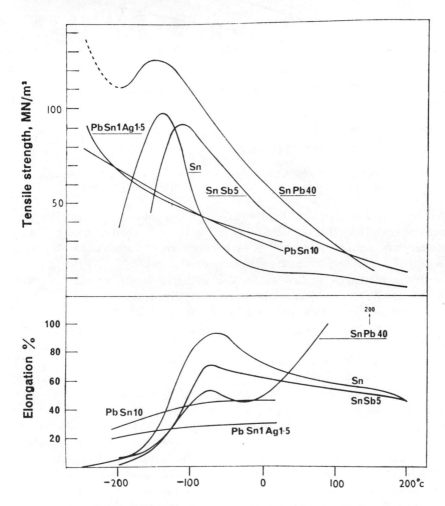

Fig. 47 – Effect of temperature on the mechanical properties of solder alloys.

Rate of testing has a significant influence on the strength of solders and soldered joints (Fig. 48).

From recent work it would appear that the mechanical strength of soldered joints and the strength of the bulk solder are not closely related. Furthermore,

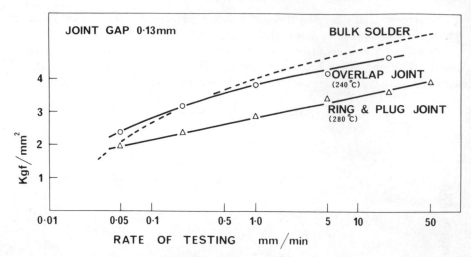

Fig. 48 – Effect of testing speed on strength of soldered joints. Ring and plug
joints made from copper; solder composition 60% tin, 40% lead.

the long-term creep strength of different alloys does not show the same order of
merit of strength as that obtained from tensile testing. In certain configurations
of electronic assemblies and in heat exchangers, fatigue failure due to repeated
reversal of stresses may be experienced in soldered joints.

Whilst the largest application for soft solders is for joining purposes, tin—
lead alloys have certain other special uses, principally because of their low melting
point and ease of casting into detailed forms. For example, tin—lead alloys
containing about 40 to 50 per cent tin are frequently used for making costume
jewellery and similar articles by centrifugal casting in rubber moulds (see Chapter
7). For increased strength, an antimonial solder alloy may often be used. In view
of the tendency for lead-containing materials to produce grey marks on the
surfaces which they contact, these articles are frequently electroplated with a
decorative coating such as a flash of gold over a copper undercoat.

The casting of tin—lead alloy sheet for organ pipe making is described in
detail in Chapter 7. These large sheets are subsequently cut to size and folded
around mandrels to form the organ pipes and the longitudinal butt joint in the
sheet is carefully soldered with the same alloy. The conical foot and the small
mouthpieces used for tuning the pipe are also attached by soldering. Clearly
these operations require skilled craftsmanship, especially for the larger pipes.

Another traditional use for solders is in the manufacture of moulds for
casting wax candles (see Chapter 4).

### 5.3 SOLDERING

Tin-based solders are used widely in the electronic, electrical and engineering
industries. For complete details of design and production of soldered joints

reference should be made to standard textbooks. However, a general discussion of the principles and processes is included here as an introduction to the subject.

Soldering is distinguished from other hot-joining processes in that the intermediate film of filler metal is an alloy of low melting point (usually containing a substantial proportion of tin) compared with the higher strength copper or copper-alloy filler metals used in brazing or silver soldering processes, or indeed with the use of the parent metal as the filler as in welding. In virtually all alloys used as soft solders, the active constituent present is tin which promotes wetting of the substrate metals, while the lead present in the commonly used binary alloys acts not only as a diluent but also has the beneficial effect of giving the alloy a melting point lower than that of the constituent metals. During the soldering process the solder wets the basis metal surfaces but a reaction at the interface between tin in the solder and the substrate metals occurs and is often considered to be necessary for good wetting to occur. Usually the reaction is evidenced in the case of copper by the formation of a thin layer of intermetallic compound $Cu_6Sn_5$ (Fig. 49) with perhaps a very thin layer of $Cu_3Sn$ present.

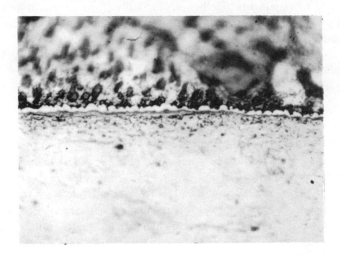

Fig. 49 — Microsection of soldered joint showing intermetallic compound layers formed by liquid—solid reaction ($Cu_3Sn$ adjacent to the copper; $Cu_6Sn_5$ adjacent to the solder). (Etched in hydrochloric acid vapour.) (Magnification × 1000.)

The ease and speed with which solder wets and bonds to a surface under given conditions are referred to as the solderability of the system, and this parameter is discussed in more detail later. A soldered joint comprises three distinct items — the basis metals requiring to be joined, the soft solder alloy, and the flux, the latter being a chemical reagent to remove surface oxides during soldering. These, together with heat to melt the solder, are the variables requiring to be controlled in the soldering process.

Good wetting by the liquid solder is the most important aspect of the soldering process and the degree of wetting which occurs depends on the free energies and the interfacial tensions at the various surfaces involved in the soldering system.

With certain imperfections in surface conditions on the basis metals, there is a pronounced increase of contact angle after initial wetting has occurred so that the solder retracts into globules, or 'dewets', leaving only an extremely thin, matte film of solder, on the surface of which these globules rest. Dewetting is undesirable since it may reduce both the quantity and the continuity of the solder in the joint and the strength of the bond.

## Requirements of a soft soldered joint

Soft soldered joints are usually required to fulfil some or all of the following functions:

(1)  To provide an electrically conductive path (for example, electronics).
(2)  To connect components together mechanically (for example, tinplate cans).
(3)  To allow heat to flow from one component to another (for example, heat exchangers).
(4)  To retain adequate strength at temperatures from cryogenic levels to well above that of boiling water (for example, freezing plant, automobile radiators).
(5)  To form a liquid- or gas-tight seal (for example, cigarette lighters).

The selection of the solder alloy and the type of flux used — and to some extent the choice of soldering technique — depend on the basis metals which constitute the parent members of the joint and the end use of the soldered joint.

Although the electrical conductivity of solder is relatively low, about 8 to 13 per cent of that of copper, the short conducting path through the solder in a joint does not normally have to be considered as adding significant electrical resistance to a circuit.

## Advantages of soldering

The major benefits to be gained from the use of soft soldering are the relative ease and speed with which it may be carried out and the versatility and range of heating techniques available. The overall cost of the process may therefore be low compared with other joining methods. The low temperatures involved in soldering will not, in general, alter the properties of the parent metals. A further advantage of soldered joints is that they may usually be taken apart by re-heating and re-made if required.

Against these factors must be placed the low mechanical strength of soldered joints.

## Making a soldered joint

The production of a soldered joint can be divided into a number of basic steps:

(1) The metal parts to be joined are shaped so that they fit together.
(2) The surfaces to be joined are carefully cleaned or prepared by pre-coating.
(3) Soldering flux is applied (the parts may sometimes be assembled at this stage).
(4) Heat and molten solder are applied and the solder usually distributes itself between the joint surfaces by capillary attraction.
(5) The joint is cooled.
(6) Flux residues are removed as required.

## Preparation of surfaces for soldering

Because wetting of the parent metals is necessary in order to produce a metallurgical bond, it is essential that there is perfect chemical cleanliness of the joint surfaces, especially where fluxes of low activity (as for electronics work) are mandatory. Cleaning is carried out by chemical or by mechanical methods.

Many basis metals may be joined in their natural state but inevitably after surface cleaning slight oxidation will occur. If strong fluxes are allowable these may be capable of dissolving these oxide films but in more critical fields, such as the electronics industry where relatively weak fluxes only are allowable, surfaces to be soldered are usually prepared by applying a coating which can impart good solderability. For example, coatings of about 8 $\mu$m thickness of pure tin or 60 per cent tin—40 per cent lead alloy will retain the highest level of solderability under most adverse storage conditions. Methods of applying tin or tin alloy coatings are described in Chapter 8.

## Soldering fluxes

Fluxes are essentially acids which have the power to dissolve or dislodge metal oxides or other films either on the substrate metals or on the molten solder itself. The most active fluxes are those based on zinc chloride or other inorganic halides which can cope with relatively severe contamination of the surfaces. Such fluxes are usually corrosive, as is the residue after soldering, so that thorough washing after completion of the joint is required.

In the fields of the electrical and electronics industries, corrosion during service or degradation of electrical resistivity of insulating materials cannot be tolerated, and therefore fluxes are mostly based on natural wood rosin which is non-corrosive. Either a rosin solution in alcohol is used as a pre-fluxing medium or the rosin may be used as a paste core in solder wire. The rosin may be increased in activity by the addition of certain other compounds, such as organic acids or halides, which enable it to dissolve oxides better and thus promote more rapid wetting by the solder. The amount of these additives is strictly limited by specifications to avoid subsequent corrosion. Table 16 shows the classification of soldering fluxes given in BS 5625.

### Table 16

Soldering flux classes in BS 5625: 1979

| BS 5625 class | Compositional basis |
|---|---|
| 1 | Zinc chloride with or without other inorganic halides and/or inorganic acids |
| 2 | Phosphoric acid and derivatives |
| 3 | Halides of organic compounds |
| 4 | Organic acids |
| 5a | Activated rosins containing halide |
| 5b | Activated rosins not containing halide |
| 6 | Non-activated rosins |
| 7 | Other organic compounds |

### Joint design

The optimum clearance in a joint is about 0.12 mm, so design of the soldered joint is important if the maximum strength is to be obtained. Normally overlap joints are recommended since the joint area may be chosen to keep the shear stress on the solder within the known limits of strength. Butt joints are inherently stronger but are susceptible to tearing with any slight non-axiality of loading. Where greater mechanical strength is essential it is preferable to obtain this by mechanical means such as a rolled seam or spot weld so that the solder as far as possible acts only as a sealant or as a path for conducting heat or electricity.

### Soft soldering processes

The techniques available for soft soldering may be classified according to the method of applying the solder, the flux and heat; brief details of these methods are given below.

#### Solder bath techniques (dip and wave soldering)

In these processes the solder bath itself acts as the heat source. Complete immersion in a molten solder bath is used for some general assembly work. Especially of interest is the soldering of solid-state electronic components to printed circuit boards (Fig. 50) in which the underside (non-component face) is brought into contact with the solder surface.

In this technique, flux is first applied to the underside of the assembly by a suitable method such as dipping into a bath, brushing or by passing it over a standing wave of foamed flux produced by aeration from a submerged porous element. The flux coating on circuit boards is then usually heated to remove the bulk of the flux solvent prior to soldering to prevent trapping of gases evolved during soldering.

Fig. 50 – A typical printed circuit assembly showing the soldered and the component surfaces.

A printed circuit assembly may be soldered by simply dipping the board flat on to the freshly cleaned surface of a solder bath maintained at 240 to 260 °C and holding it there for a few seconds. Drag soldering consists of moving the board along the surface of an elongated bath (Fig. 51).

In wave soldering, solder is pumped out of a narrow slot to produce a standing wave against the crest of which the circuit board passes on a conveyor (Fig. 52). The purpose of the wave is to renew the solder surface continually and thus present a virtually oxide-free surface to the board.

A bath of solder is also used for soldering the side seams of tinplate cans, but in this case a cylindrical steel roll rotates partly immersed in the solder bath so that it becomes coated with molten solder. As the can bodies move at high speed along the length of the roll with their side seams in contact with it, the solder is transferred from the roll to the seam (Fig. 53).

### Soldering irons

The use of a soldering iron either manually or in automatic equipment is the most widely used method of soldering. The bit of the iron stores and carries heat from the heat source and delivers molten solder and often flux to the work.

Bits are usually made from copper because this metal combines good wetting properties with the best thermal capacity and thermal conductivity. Since plain copper bits dissolve and erode rather quickly in molten solder they are often plated with a thick coating of iron or nickel which is attacked much more slowly and ensures that the shape of the bit, essential for good soldering practice, is maintained without the need for regular reshaping.

The size and shape of the bit are largely determined by the specific joint configuration, controlling the amount of heat that has to be supplied during each jointing operation. The heat input − wattage in the case of an electric iron − is determined by the rate of working, i.e. the rate at which heat must be supplied to maintain the bit at the required temperature. The majority of soldering irons use built-in electrical heaters and may incorporate a thermostat or electrical energy controller. Gas-heated irons are used for some applications.

### Hotplate soldering

During soldering on a hotplate, jigs are necessary to hold the components rigidly in their correct positions. Solder may be hand fed, or preformed shapes or paste can be used. The assembly must be carefully removed from the hotplate for cooling to avoid vibration while the solder is still partially liquid.

### Oven soldering

The conditions during oven soldering are somewhat similar to those for hotplate soldering, since jigging and the use of preforms of solder are essential features. Heating is much slower because heat transfer is largely by radiation.

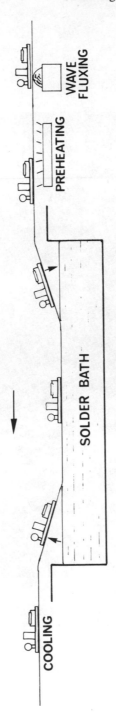

Fig. 51 – Diagram illustrating principle of drag-soldering.

Fig. 52 — Mass soldering of printed circuit board assemblies using a wave soldering machine. The photograph shows a loaded board passing from the flux drying section to enter the solder wave. It takes only a few seconds for all the component leads to be soldered to the tracks on the underside of the board.

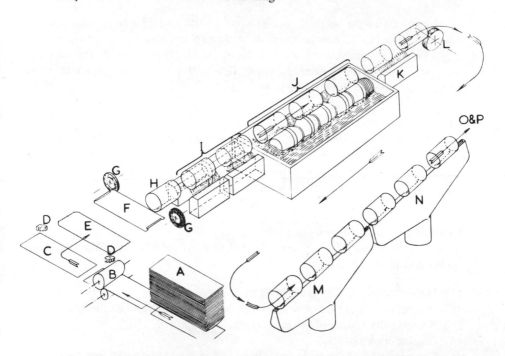

Fig. 53 – Diagram showing manufacturing stages for three-piece tinplate cans including the roll-soldering machine for the side seams. Key: A: Tinplate blanks. B: Fluxing station. C & D: Edge knurling. E: Notching. F: Hook forming. G: Fluxing with brushes. H: Body forming and lock seaming. I: Pre-heat. J: Soldering. K: Post-heat. L: Wiping to remove excess solder. M & N: Forced air cooling. O & P: Delivery to end flanging machine.

*Radio frequency induction soldering*
This method is relatively expensive in terms of capital equipment costs; moreover, correct design of the induction coil for each particular assembly is essential. The rate of heating is pre-set by a power control to give a heating time of several seconds. Solder washers or preforms are assembled with the components being joined and the surfaces of the components should have a high level of solderability to ensure good wetting and flow of the solder. The components are coated with flux solution prior to assembly.

*Electrical resistance (conduction) soldering*
The work-piece is connected to one terminal of a low-voltage transformer while a carbon rod in a holder is connected to the other terminal so that, at the point of contact of the carbon with the work-piece, contact resistance heating occurs and flux and solder can be applied in the usual manner.

In an alternative method, the area on which soldering is to take place is heated by passing an electric current between electrodes placed on either side of it. This technique is often used in the electronics field for soldering flat-pack integrated circuits on to printed circuits. The solder is present as thick coatings of tin—lead alloy on the surfaces to be joined.

*Ultrasonic soldering*
An ultrasonic transducer in a solder bath can be used to pre-tin electrical component terminations without flux. An electric soldering iron incorporating a transducer has been marketed for the soldering of aluminium; the vibrations rupture the alumina skin and allow the underlying metal to be wetted.

**Reliability of soldered joints in service**
To obtain the required high level of reliability in soldered joints the following points should be observed:

(1)  Good mechanical design of the joints.
(2)  Correct choice of surface preparation, surface coating and its thickness.
(3)  Solderability testing of the components to be joined.
(4)  Correct choice of flux.
(5)  Consideration of the optimum method of applying the solder and heat.

All of the soldering processes outlined in the previous section will provide a joint of good reliability provided that the parameters in the soldering process are correctly controlled and, of more importance, that the initial preparation stages before the final soldering operation are controlled very closely. With the correct choice of surface cleaning treatment, or alternatively a suitable metallic coating applied over the clean basis metal, the soldering operation presents no difficulty. However, in order to achieve the maximum possibility of achieving good wetting, it is advisable, for electronics work, to carry out solderability tests on the components to be joined. Several different test methods are available, a common feature being a measurement of the time required to wet the given surfaces with solder under closely specified conditions.

Since there is no suitable non-destructive physical test for the quality of soldered joints, inspection of completed soldered assemblies from the production line should be made using, for example, a magnifier of not more than $\times 10$ power. The 'pass' criteria are that soldered connections should be shiny and should have smooth, concave, low contact angle fillets at the component surfaces (Fig. 54a). A sharp line of demarcation at either end of the fillet (high contact angles) may be indicative of inferior wetting by the solder. Unwetted areas, large cavities in the joint and dewetting of the solder are all possible reasons for rejection (Fig. 54b).

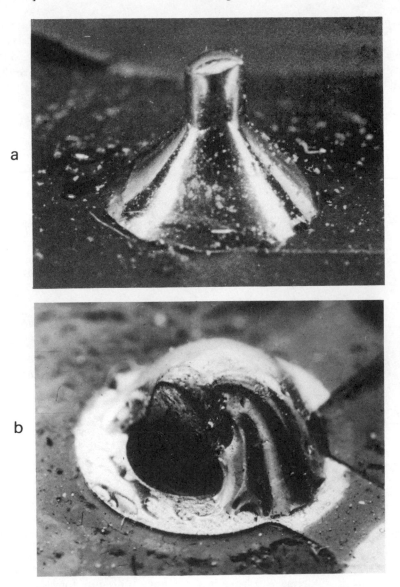

Fig. 54 – Appearance of satisfactory soldered joint (a) and one of inadequate
quality showing poor wetting of the components (b) magnified.

Environmental conditions during service may also affect the life of a soldered
joint. To some extent allowance may be made for these conditions, for example
by choosing solder alloys of higher melting point to provide resistance to creep
under stress at service temperatures above 100 °C.

A phenomenon associated with elevated service temperatures (for example over 100 °C) is the growth of layers of intermetallic compound at the interface between the solder and the basis metals by a diffusion reaction (Fig. 55). Ultimately all the tin in the solder may be consumed and this may cause the joint to become brittle or lose strength.

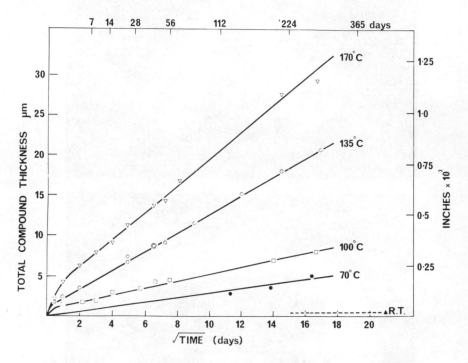

Fig. 55 – Rate of growth of intermetallic compound by solid state diffusion at various temperatures in tinplated copper.

Soldering has the merits of providing a relatively strong metallurgical bond while it requires heating only to relatively low temperatures. Although some other mechanical connection techniques may be applicable for general sheet-metal work, soldering is still the major jointing procedure for electronics assemblies.

## FURTHER READING

1.  *Soft-Soldering Handbook,* International Tin Research Institute Publication No. 533.
2.  Allen, B. M., *Soldering Handbook,* Iliffe, London, 1969.

3. Manko, H. H., *Solders and Soldering,* McGraw Hill, New York, 1964.
4. *Soldering Manual,* American Welding Society, New York, 2nd edn., 1979.
5. Thwaites, C. J., The attainment of reliability in modern soldering techniques for electronic assemblies, *International Metallurgical Review* No. 166, Vol. 17, p. 149, 1972.
6. Strauss, R., *Das Löten für den Praktiker,* Franzis-Verlag, Munchen, 1978.
7. Thwaites, C. J. and Barry, B. T. K., *Engineering Design Guide 07 Soldering,* Oxford University Press, 1975.
8. *Photographic Guide to Soldering Quality,* International Tin Research Institute Publication No. 555.

# 6

# Bearing Alloys

The US Patent No. 1252, granted to Isaac Babbitt in 1839, described the use of tin-base whitemetal bearings backed by a stronger shell, but Leonardo da Vinci seems to have published similar ideas about two centuries earlier. Babbitt stated that the 'boxes' (which would now be called the shells or backing) can be made of any kind of metal which has sufficient strength and which is capable of being tinned. He then indicated that the 'boxes' are lined with tin alloy which derives hardness by virtue of the addition of antimony and copper. Thus Babbitt described the need for a strong backing, adequate bonding of the whitemetal and a lining of tin-base alloy in which antimony and copper were the principal alloying elements. The alloy composition suggested was close to that presently widely used, namely tin with 7 per cent antimony and 3 per cent copper.

**Requirements in bearing alloys**
Properties that are required of a good bearing metal include:

(1)  Surface properties that resist seizure by maintaining a lubricant film.
(2)  High strength to resist deformation by both steady  and fluctuating loads.
(3)  Sufficient hardness to combat wear.
(4)  Good conformability to compensate for imperfections in shaft geometry.
(5)  Good embeddability to absorb dirt and debris.
(6)  High corrosion resistance.

Some of these requirements are conflicting because (2) and (3) involve hardness while (4) and (5) require softness. A bearing alloy therefore necessarily represents a compromise in its properties and conventional fatigue tests on the alloy itself will only give a very general guide to bearing performance and will not necessarily be related to behaviour of a thin-walled bearing. Several types of test units have been designed and built by commercial bearing manufacturers which attempt to simulate the real service conditions of an engine, especially in automotive applications.

In a lubricated journal bearing, fluid lubrication results from the formation of an oil wedge between the bearing and the journal surfaces, and the resultant pressure tends to 'float' the journal from the bearing. In a perfectly designed bearing, the continuous oil film formed in this way would separate the bearing surfaces and prevent metallic contact. In practice, however, departures from ideal conditions, such as foreign particles suspended in the oil or irregularities in the metal surfaces, cause local or general breakdown of the oil film and result in 'boundary' lubrication, which allows momentary local welds to form, which can, in turn, lead to seizure.

Bearing metals based on tin do not, in themselves, provide a low coefficient of friction unless at least trace amounts of lubricant are present. It is therefore perhaps the chemical nature of tin which ensures good adhesion of films of oil resulting in the excellent performance of a tin-containing bearing alloy. As stated previously, no tin alloy has great mechanical strength, so that compositional changes in tin-rich alloys have only a limited benefit. To increase the load-carrying capacity, bearings are almost always made by bonding the whitemetal to a stronger backing such as bronze, steel or cast iron (Fig. 56). One effect of the backing is to prevent flexing of the whitemetal, thus reducing the stresses which are responsible for fatigue. The extent to which the backing is capable of supporting stresses at the whitemetal surface depends on the thickness of the whitemetal. The relationship between bearing life and whitemetal thickness in a series of accelerated tests is shown in Fig. 57.

Shell bearings with a whitemetal thickness down to 0.05 mm have been used where maximum resistance to fatigue is essential and where precision machining and assembly are possible. The use of these 'micro-bearings' is confined to the automobile industry where fine limits in machining allow bearings to be inserted directly to give the necessary clearances and alignment. They give freedom from fatigue cracking at loads almost twice as high as can be employed with the normal thicker linings. Obviously their main disadvantages are reductions in embeddability and conformability.

The main categories of bearing alloys containing tin are whitemetals, aluminium—tin alloys and bronzes. Whereas whitemetals have to be used as a lining on a shell or backing which provides mechanical support, both of the other two categories include alloys used as solid bearings without backings.

## Metallurgical considerations for bearing metals

Apart from the required properties of bearings listed earlier, namely adequate hardness combined with conformability and, of course, fatigue strength, a metallurgical requirement still regarded perhaps as essential to a good bearing alloy is that of having a two-phase structure. In tin-based whitemetals hard intermetallic particles are present in a soft matrix. The argument usually advanced is that the hard particles form a 'pavement' which supports the load while the soft matrix is worn down to a level slightly below that of the 'pavement' and

Fig. 56 – Fitting the crankshaft of a large modern diesel engine into the tin-rich whitemetal bearing housings. The cap halves of the bearings are seen in the foreground. Whitemetals provide the necessary combination of properties in this type of application.

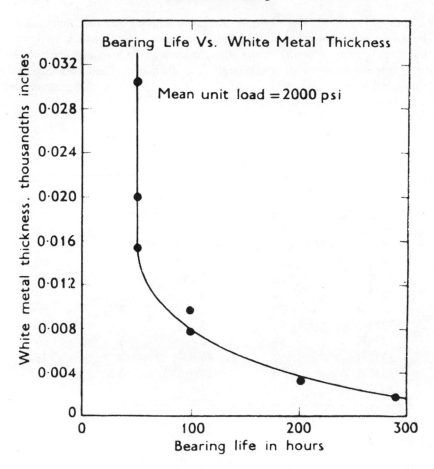

Fig. 57 − Influence of whitemetal thickness on fatigue strength of bearings.

thus provides a series of channels for supply of lubricant. This hypothesis is now a matter of controversy since, for example, the aluminium−tin alloys, which have been well proven industrially, in fact have a relatively hard matrix containing many discrete islands of a soft phase (tin). The significant nature of the duplex structure is concerned with performance, as has been shown in bronzes where variations in the number and size of the particles of hard phase alter the wear resistance markedly.

## 6.1 WHITEMETAL BEARINGS

Whitemetals can conveniently be subdivided into three types − high-tin alloys, high-lead alloys and intermediate alloys which contain significant proportions of both tin and lead. For all three the major alloying and hardening addition elements are antimony and copper. Thus the dispersed intermetallic compounds found in the microstructure of the alloys in each group are similar; where they differ is in the composition and properties of the matrix. Table 17 indicates the whitemetal bearing alloys given in BS 3332: 1961, whilst Table 18 suggests areas of application for different bearing alloys.

**Table 17**

Whitemetal bearing alloys (BS 3332:1961)

| Alloy grades | Nominal composition (%) | | | |
|---|---|---|---|---|
| | Sn | Sb | Cu | Pb |
| 3332/1 | rem. | 7.5 | 3.25 | <0.35 |
| 3332/2 | rem. | 9.25 | 4.25 | <0.35 |
| 3332/3 | >80 | 10 | 5 | 4 |
| 3332/4 | 75 | 12 | 3.5 | rem. |
| 3332/5 | 75 | 7 | 3 | rem. |
| 3332/6 | 59 | 10 | 3 | rem. |
| 3332/7 | 12 | 13.5 | 0.75 | >75 |
| 3332/8 | 5.5 | 15.5 | <0.5 | rem. |
| 3332/9 | 68.5 | – | 1.5 | –[a] |

a rem. Zn

*Tin-based alloys*
These are all based on the tin–antimony–copper system, the antimony content being up to about 12 per cent and the copper up to 10 per cent in commercially used alloys. When cast, the alloys are in a metastable state and the phases present are indicated in Fig. 58.

Because the matrix is a supersaturated solution of antimony in tin the alloys tend to approach equilibrium in use by precipitation of fine particles of SbSn, particularly when run under hot conditions, but the resulting changes in structure are not considered to be of practical importance. The commercial bearing alloys are selected from the phase fields A and B. In the latter (i.e. for antimony contents up to about 8 per cent) primary crystals of $Cu_6Sn_5$ separate as needles in a random array through the matrix consisting of a divorced eutectic of tin-rich solid solution plus $Cu_6Sn_5$, while in field B a secondary separation of the cubes of SbSn occurs in addition to $Cu_6Sn_5$ (Fig. 59).

The SbSn cubes are less dense than the liquid from which they separate and,

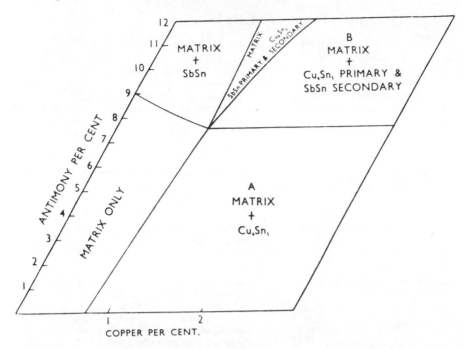

Fig. 58 – A corner of the tin–antimony–copper ternary equilibrium diagram relevant to whitemetal bearing alloys.

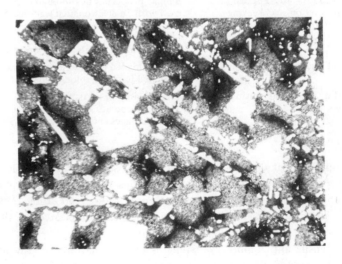

Fig. 59 – Microstructure of cast tin, 9% antimony, 4% copper alloy showing SnSn cuboids and $Cu_6Sn_5$ needles in a tin-rich matrix. (Etched in 2% Nital.) Magnification ×100.)

## Table 18

### Typical applications for bearing alloys

| Bearing type | Application areas |
| --- | --- |
| Tin-rich whitemetal (> 0.3 mm thick) cast on steel, cast iron or bronze backings | Large marine Diesel engine main and connecting-rod bearings; ship propeller shaft bushes; steam turbines and large generating equipment bearings and thrust pads; earth-moving machinery; railway axle-boxes; for unhardened journals and where maximum embeddability and conformability are necessary. |
| Tin-rich whitemetal (< 0.3 mm thick) on steel backings | Shell bearings for small and medium internal size combustion engines; compressors; general machinery; unhardened journals allowable. |
| Lead-rich whitemetal (on backing) | For less exacting service conditions in applications similar to those for tin-rich whitemetals; automotive engines (USA); to effect a cost saving. |
| Aluminium–tin (20–40% Sn) on steel backings | Gas turbines; aircraft landing gear bushes; small and medium internal size combustion engine big end and main bearings where the loading is too high to allow use of whitemetal and unhardened journals are in use (20% Sn alloy); marine Diesel crosshead bearings (40% Sn alloy). |
| Aluminium–tin (6% Sn) solid | High-speed aircraft engines; machinery; pumps; greater fatigue strength than the higher Sn content alloys; requires a hardened journal. |
| Copper–lead (10–40% Pb) cast or sintered powder on steel backing (with or without overlay plating of lead–10% tin, or lead–7% indium) | Similar properties to aluminium–20% tin when used with an electroplated lead-rich overlay; machinery with hardened journals required if no overlay used, and where risk of corrosion by hot oil is absent; high fatigue strength. |
| Bronze and leaded bronze, solid | Heavily loaded and low speed machinery using hardened journals; rolling mills; bridge bearings. |
| Graphited or plastic impregnated tin bronzes | Unlubricated or partially lubricated operation with high operating temperatures and low loads; locks, barrages and hydroelectric plants. |

under conditions of slow solidification, tend to segregate. The classical theory is that the primary separation of $Cu_6Sn_5$ needles forms a network in which the SbSn cubes are entangled and thus prevented from segregating. There is also evidence that the separation of the SbSn cubes is nucleated by the primary $Cu_6Sn_5$ and that they actually grow around it.

The tensile and compressive strengths and the hardness of tin–antimony–copper alloys generally increase as the antimony and copper contents are raised. The more highly alloyed materials (especially with 8 to 10 per cent copper) have very low ductility.

The fatigue strength of these bearing alloys is perhaps of more importance than their tensile properties. Generally fatigue strength increases with antimony content, most rapidly when it is all in solid solution (i.e. up to about 8 per cent) and more slowly above this level since the presence of SbSn cubes in the microstructure has less strengthening effect (Fig. 60).

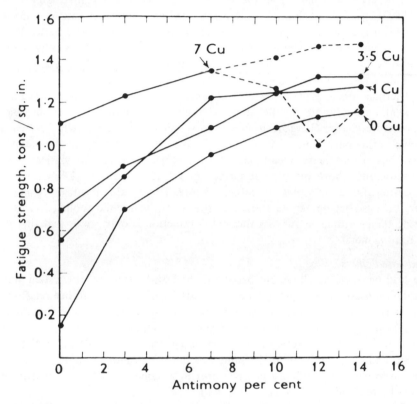

Fig. 60 – Effect of antimony on fatigue strength of tin–antimony–copper alloys.

For a constant antimony content, the addition of copper up to about 1 per cent forms a fine eutectic and brings about a marked increase in fatigue strength, especially in alloys of low antimony content. Further increases in copper content result in the formation, under normal casting conditions, of relatively large needles of primary $Cu_6Sn_5$ and these have much less effect on the fatigue strength than the eutectic $Cu_6Sn_5$.

Regarding the effect of other alloy additions on fatigue strength of white-metal alloys, it has been claimed that small additions of bismuth, tellurium or cadmium are beneficial and, for more severe loading conditions, some manufacturers regularly supply alloys containing about 1 per cent cadmium. Other elements such as silver, nickel and, more recently, beryllium and chromium have been added as structure-refining or nucleating agents and in this manner raise the fatigue strength significantly.

The cracking which can eventually occur in whitemetal bearings subjected to large cyclic loads is generally accepted as being due to fatigue. Typical cracks take a perpendicular path from the surface of the bearing towards the backing and, having penetrated almost to the bond, continue parallel to, but just above it. In advanced stages of fatigue the surface of the bearing shows a network of cracks which penetrate almost to the bond and spread laterally, so that large pieces of whitemetal become detached. There is little or no evidence that fatigue cracks start at the bond, even in those cases where the bond is weak.

Thermal effects may also cause problems in coarse-grained, tin-rich bearing alloys because crystals of the tin—antimony solid solution are anisotropic. Temperature changes have been shown to set up micro-stresses which, after a number of thermal cycles, result in creep of those crystals having suitable orientations. This in turn causes measurable steps to appear at grain boundaries; those grains standing above the general surface may lead to local breakdown of the lubricating film and 'wiping' or fatigue failure. The backing itself, especially if large, can also set up tensile stresses in the whitemetal while cooling from the casting temperature, because its thermal contraction is only about half that of the bearing metal.

*Lead-based alloys*
The lead-rich bearing alloys are based on the lead—antimony—tin system and, like the tin-based alloys, have a structure consisting of hard crystals in a relatively soft matrix. The alloys in commercial use fall into two classes — those containing 12 to 18 per cent antimony and up to 5 per cent tin which have primary crystals of antimony-rich solid solution in a matrix of the ternary eutectic lead—antimony —SbSn, and those with about the same amount of antimony and 10 to 12 per cent tin in which the primary phase is present as cubes of SbSn in a matrix of pseudo-binary lead—SbSn eutectic (Fig. 61).

The second class of alloy with the higher tin content has generally better properties than the first, although similar solidus temperatures. All the lead-based alloys are prone to segregation during casting as both the antimony-rich and the SbSn primary crystals are less dense than the high-lead matrix and tend to float. Segregation is usually controlled by the addition of small amounts of copper.

Lead-based alloys are generally regarded as a cheaper substitute for tin-based alloys and soften more rapidly than tin-based alloys at elevated temperatures.

*Tin—lead-based alloys*
Intermediate between the tin-rich and lead-rich whitemetals is a whole range of alloys which contain both tin and lead in the matrix. These alloys are much less used industrially.

Fig. 61 — Microstructure of lead, 13% antimony, 12% tin, 1% copper alloy showing SbSn cuboids in a pseudo-binary lead–SbSn matrix. (Etched in 2% Nital.) (Magnification × 100.)

At room temperature the mechanical properties of the intermediate alloys are not greatly different from those of the high-tin and high-lead alloys, but there is a gradual rise in hardness and tensile strength as the amount of tin is increased. However the resistance to compression and to fatigue is rather lower for the alloys containing from 20 to 50 per cent tin and at elevated temperatures the intermediate alloys are markedly inferior as regards mechanical properties because the matrix contains a peritectic complex which melts at about 180 °C compared with a solidus at about 230 to 240 °C for the high-tin and high-lead alloys.

The intermediate alloys have an unquestioned advantage over the high-lead alloys in their greater ease of casting and they are less prone to segregation.

### Manufacture of whitemetal bearings

The bearing alloy is cast on to the carefully tinned backing so that a true metallurgical bond is obtained. This is probably the most important part of the whole manufacturing process.

Essentially the pre-tinning of new bearing shells is described under hot-tinning procedures in Chapter 8. After suitable surface preparation, the shells are dipped in an aqueous zinc-chloride flux and lowered into the bath of molten tin at about 300 °C for ferrous shells or 260 °C for bronze backings. A few minutes' immersion is often required to bring the shell up to tinning temperature and cast iron shells benefit from immersion times up to 20 minutes. Bronze shells require the minimum acceptable immersion time and minimum tinning temperature in order to avoid an excessively thick layer of intermetallic compound forming on their surfaces, which can reduce the strength of the bond to the whitemetal cast on to them at some subsequent time.

Very large bearing shells cannot usually be dipped into a tinning bath and are generally pre-heated and tinned by a manual wiping procedure in which flux is applied and a stick of tin is melted on to the surface and wire-brushed all over to give a uniform tin coating.

Assuming that the shell has been properly tinned, good adhesion between the backing and lining depends primarily upon two factors. Firstly, the tin layer on the backing must be completely molten when the whitemetal is poured on to it, so that any oxide films are floated off by the alloy. Secondly, the whitemetal must solidify from the bond inwards to prevent the formation of solidification contraction cavities at the bond which would reduce bond strength. When an adequate supply of liquid metal is available for feeding, the contraction cavities develop in the last part of the metal to solidify, i.e. at the surface, and any porous metal is removed when the bearing surface is machined. The rate of cooling also affects the bond strength but to a lesser extent, and the optimum rate depends on the composition and temperature of the whitemetal and on the size of the shell to be lined.

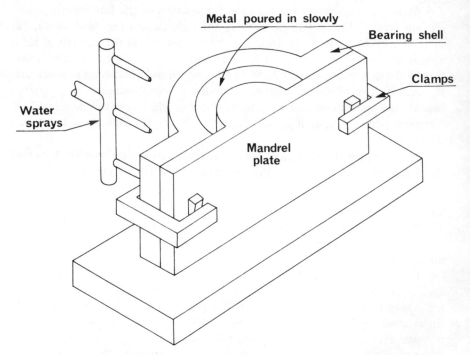

Fig. 62 — Diagram showing arrangement for casting whitemetal on to a pre-tinned cylindrical half bearing shell.

Fig. 63 — Equipment for centrifugally casting whitemetal on to large bearing shells.

In the usual lining practice, a semi-cylindrical pre-tinned bearing shell is set up with its axis vertical against a mould plate. A semicircular mandrel is inserted between the plate and the bearing shell in order to leave a uniform gap into which the bearing alloy is poured. The complete casting jig preferably stands on an insulating surface to reduce heat losses (Fig. 62). The metal at about 50 °C above its liquidus is poured in from the top and when the cavity is filled, water sprays or air jets are directed at the exterior of the bearing shell to induce directional cooling. Thrust pads are cast normally on their backs, often in a shallow tray which is filled with water after casting to cool the backs of the pads.

Centrifugal casting often provides a casting free from porosity and the shape and nature of many bearing bushes make them ideal subjects for this technique. This method of lining is frequently applied to bearing sleeves of all sizes up to about 1 m in diameter (Fig. 63). In tin-base alloys under the influence of centri-fugal force, the primary $Cu_6Sn_5$ needles which separate move away from, and the SbSn cubes move towards, the centre of rotation. The degree of segregation, which depends on centrifugal force, is a function of the speed of rotation, and speeds at which satisfactory structures are obtained must be determined empirically. The temperature of pouring, the temperature of the shell prior to lining and the rate of cooling also affect the rate of solidification and hence the amount of segregation.

In the case of lead-base alloys, the higher specific gravity of the liquid matrix causes the segregation pattern to differ from that produced in tin-rich alloys, and in this respect careful centrifugal lining will give a better distribution of the suspended intermetallic compound crystals than would a static casting procedure.

Strip material for stamping small automotive steel-backed shell bearings is made by passing a continuous coil of narrow steel strip successively through a surface preparation unit, a bath of molten tin and immediately into a casting channel in which molten whitemetal is continuously metered on to the moving tinned steel between edge 'dams' and is solidified by water sprays applied to the underside of the strip (Fig. 64). The whitemetalled strip is cut into pieces which are formed into a semi-cylinder and finally machined to close tolerances ready for fitting into engines.

## 6.2 ALUMINIUM–TIN BEARINGS

In the 1930s higher strength alloys of aluminium were introduced for aircraft engines. These were aluminium–6 per cent tin alloys and were used as solid cast bearing materials, normally without a steel backing. Additional strength is imparted by small additions of copper, nickel, silicon or magnesium combined with a heat treatment. However, although these low-tin alloys exhibit good fatigue properties, seizure resistance and reasonable embeddability, they tend to

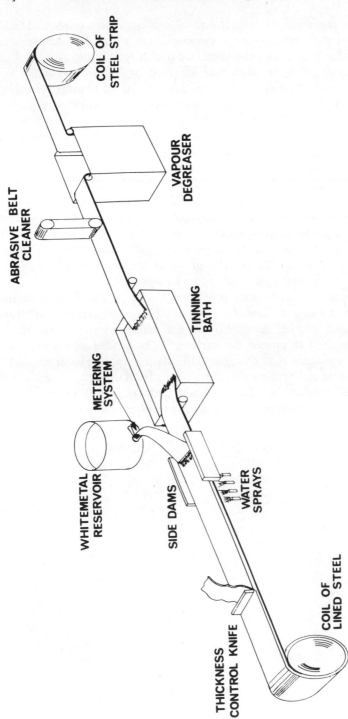

Fig. 64 — Diagram of continuous line for pre-tinning and casting whitemetal on to steel strip for cutting into pieces to be formed into bearing half-shells for automotive applications.

have a low conformability and to have a high coefficient of thermal expansion. This means that relatively large clearances are required which renders them unsuitable for most automotive engines. For this application, in order to combine a fatigue strength better than that afforded by whitemetals with the other requirements of a bearing, aluminium—20 per cent tin bearings were developed in the 1950s. These have a lower fatigue strength than the 6 per cent tin alloys, but better conformability, and can be used with unhardened shafts. Aluminium alloys with 30 and 40 per cent tin are also produced commercially for certain areas of application. Where these aluminium—tin alloys are employed, they are always used as shell bearings with a steel backing to which they are bonded.

The aluminium—20 per cent tin alloy in cast form contains complete intergranular films of tin giving a low-strength material (Fig. 65).

This structure is broken down by cold rolling to form longitudinal stringers of tin and a subsequent annealing treatment for about 1 hour at 350 °C (i.e. above the melting point of the tin phase) partially spheroidises the tin to give a 'reticular' network of tin particles in the aluminium matrix (Fig. 66), which usually contains about 1 per cent copper to solution-harden it.

A continuous roll-bonding procedure is used to attach the aluminium—tin alloy strip to a steel backing. As a prerequisite, an interlayer of pure aluminium is roll-bonded to either the steel or the aluminium—tin strip using a rotary wire-brushing method to prepare the surfaces to be bonded. If the alloy is in the fully hard condition after rolling and the steel is soft, the hardening of the steel backing during roll-bonding reduction is small and does not impede subsequent bearing-shell production. Finally a heat treatment is given to the composite to improve the bond strength. This type of bearing is mass produced for use in the

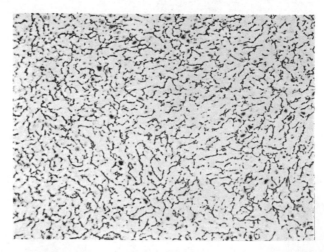

Fig. 65 – Microstructure of cast aluminium—20% tin alloy showing intergranular films of tin. (Etched in 2% Nital.) (Magnification ×150.)

Fig. 66 — Microstructure of aluminium–20% tin alloy after cold rolling, roll-bonding to steel with an interlayer of pure aluminium and final heat treatment to spheroidise the tin particles. (Etched in 2% Nital.) (Magnification × 100.)

automotive industry and has now virtually entirely replaced whitemetal bearings because the high power-to-weight ratio required in modern engines results in high bearing loadings.

The major problem with these aluminium—tin alloys is that, although there is complete liquid solubility of tin in molten aluminium, the solubility of tin is virtually zero below the melting point of aluminium. Thus, when an aluminium—tin alloy is cast from above the melting point of aluminium, primary crystals of aluminium solidify from the melt to leave an intergranular network of liquid tin which eventually freezes at 232 °C. The castings thus produced are therefore hot-short unless care is taken to ensure that all liquid tin at grain boundaries has solidified, that is that the temperature is below 232 °C.

The usual precautions are taken to degas the aluminium with chlorine gas or with solid pellets to remove hydrogen from solution in the metal before tin is added to the melt.

Other aluminium—tin alloys may contain up to 40 per cent tin and perhaps 1 per cent copper which remains in solid solution in the aluminium primary crystals. The alloy may be chill cast in steel moulds and may also be continuously cast using water-cooled graphite dies. The conditions for the latter are obviously critical in order to ensure complete solidification of the tin phase before the cast rod leaves the die.

The alloys containing about 6 per cent tin are easier to cast and are normally used in the as-cast condition. Those containing 20 to 40 per cent tin are rolled into sheet, which is subsequently bonded by rolling on to a steel backing in order to produce thin-wall shell bearings. For the latter process, the cast material is cold

Fig. 67 – A selection of sintered bronze powder components for automotive and other applications.

rolled to break up the intergranular envelopes of tin surrounding the aluminium grains and after rolling it is heat treated to cause spheroidisation of the tin particles. This produces a structure which is more suitable as a bearing material because the tin particles are randomly dispersed throughout the aluminium matrix.


## 6.3 BRONZE BEARINGS

Phosphor bronze bearings containing typically 10 per cent tin and about 0.5 per cent phosphorus are very strong and suitable for heavy load, high temperature applications such as rolling mill bearings. However, many bearing materials in this category contain lead (leaded bronzes), which has the effect of improving their surface properties, such as conformability, at the expense of strength and resistance to wear and deformation. This is due to the lead being present as discrete spherical particles in the structure, owing to the non-solubility of lead in copper. A leaded bronze with a composition copper—5 per cent tin—20 per cent lead has excellent bearing properties and is capable of operating under conditions of poor lubrication; it has a wide range of uses in railway and agricultural machinery bearings and in certain internal combustion engine bearings. In comparison to copper—10 per cent tin—10 per cent lead alloy, it has better surface properties but reduced strength and hardness.

Sintered-bronze bearings made by pressing atomised bronze powder into the required shape and sintering by holding at an elevated temperature (for example 800 °C) contain many small interstices which will accept lubricants into the structure and hence can be pre-impregnated with oil. These bearings (Fig. 67) are widely used in small machinery because they eliminate the need for regular maintenance. Sintered bronze or copper—lead bearings with a steel backing for use as shell bearings are made by a continuous strip process.

'Dry' bearings are designed to operate without regular lubrication or with minimal lubrication. They normally have a low load-carrying capacity but are particularly useful in harsh environments such as high and low temperatures, where lubricating oil would not function effectively, or else in applications where the presence of a lubricating oil would act as a contaminant. These bearings comprise a porous sintered-bronze lining bonded to a steel backing, which is impregnated with PTFE or other plastics material.


## FURTHER READING

1.  *Babbitt Alloys for Plain Bearings,* International Tin Research Institute Publication No. 149.
2.  Thwaites, C. J., Developments in plain bearing technology, *Tribologia e Lubrificazione,* Vol. 10, p. 94, 1975. (International Tin Research Institute Publication No. 513).

3. Pratt, G. C., Materials for plain bearings, *International Metallurgical Review* No. 174, Vol. 18, p. 62, 1973.
4. Neale, M. J., *Tribology Handbook,* Butterworth, London, 1973.
5. *Metals Handbook,* 8th edn., Vol. 1, Properties and Selection of Metals, American Society for Metals, 1961.
6. *Aluminium—Tin Alloy Bearings,* International Tin Research Institute Publication No. 463.
7. West, E. G., *Copper and its Alloys,* Ellis Horwood Ltd., (this series), 1982.

# 7

# Melting, Casting and Fabrication of Tin and Tin Alloys

This chapter is devoted principally to the practice of melting, casting and fabrication for alloys of high tin content, namely soft solders, pewter, bearing metals and die-casting alloys. However, there are many other alloys where tin is a minor constituent alloyed to the extent of perhaps 5 to 20 per cent with different basis metals. Details on the melting, casting and fabrication techniques for these materials should be obtained from books concerned with those particular basis metals. For example, full details regarding tin bronzes are best obtained from books concerned with casting and fabrication of copper alloys and hence only a relatively brief account will be given in the present chapter.

The reason for this division is obvious in that generally tin-rich alloys melt at low temperatures, for example 300 °C or below, and present few problems regarding melting and casting, whereas tin bronzes, for example, require melting temperatures of about 1200 °C, so that such problems as refractory materials for the crucibles, melt control and oxidation or slagging problems are totally different from those existing with tin alloys of low-melting-point.

As already indicated in Chapter 1, tin has a combination of rather special properties which include low melting point, high fluidity when molten, negligible gas solubility, readiness to form alloys with other metals, softness and hence good formability. The metal is non-toxic and has a high boiling point and therefore no losses by volatilisation occur during melting.

The main applications of alloys of high-tin content as finished castings are pewter, type metal, bearing alloys and solders. Pewter usually contains at least 90 per cent tin strengthened with small amounts of antimony and copper; it is used for domestic and decorative articles. Pewter or other alloys such as soft solders are also used to manufacture inexpensive costume jewellery and miniature figures. Cast type metal is a lead-rich alloy containing a low percentage of tin. Bearing metals are tin—antimony—copper or tin—lead—antimony—copper alloys, usually cast on to supporting backings. Solders, which are in the main tin—lead alloys, are usually first cast into billets, sticks, etc., which subsequently may be fabricated into other suitable forms for use by industry. All these applications of tin-containing alloys are described in detail in Chapters 4 and 5.

## 7.1 MELTING

The melting operation is particularly easy, due to the low melting point of tin, which enables tin and tin-rich alloys to be melted under conditions which do not normally produce contamination either from the crucible or from the atmosphere.

The equipment required for melting tin and tin alloys is relatively simple and any convenient form of heating of the crucible may be used. The crucible or melting bath must be provided with a temperature controller to prevent over-heating. For making up alloys the crucible can be of a fireclay material but more often is a hemispherical cast iron pot or welded mild-steel vessel of rectangular or cylindrical shape, fitted permanently into the heating chamber. Metal crucibles are strong and robust enough to contain a very large mass of molten metal with safety. Attack on iron or steel vessels by molten tin and the consequent pick-up of iron by the melt are not problems provided the temperature is controlled at the recommended level, which does not usually exceed 50 °C above the liquidus of the alloy being made. Additionally, the surface of the cast iron or steel should preferably be treated to prevent wetting by the tin; the pot may be intentionally oxidised (blued) by heating to about 500 °C in steam or after applying a light film of oil, or alternatively it may be coated with iron oxide paint or with a chalk or lime wash. In the rare cases where a strong flux such as fused zinc chloride is present on the molten tin surface to aid in the making up of some alloys, some wetting of the crucible will take place, accompanied by the formation at the interface of the compound $FeSn_2$. However, after this initial compound layer has formed, the rate of attack is extremely low at the relatively low temperatures employed.

Molten tin does not dissolve oxygen or nitrogen from the air or hydrogen from a gas furnace atmosphere and the surface tin oxide layer formed when the metal is held molten in an oxidising atmosphere is fairly impermeable to oxygen, thus providing some measure of protection against further oxidation of the underlying molten metal. Because of this, tin and tin-rich alloys are usually melted in air without the use of a blanket cover of flux and, provided the metal is not grossly overheated, the loss due to oxidation is small. A fume extraction hood over the melting pot would be normal and is particularly important when alloys of high lead content and those containing cadmium are to be made up or melted.

If the melt in a metal pot is allowed to solidify, care is required when remelting. The method of heating must be such as to ensure that the solidified metal melts at the sides before the bottom, to avoid pressure being built up by the expansion which occurs during fusion, and the consequent violent ejection of molten metal, perhaps with distortion of the bath itself.

Refractory fireclay or plumbago crucibles are used for small melts or for making alloys of special composition. However, they must not be used in conjunction with fluxes containing hygroscopic materials such as zinc chloride, since the flux is absorbed by the refractory and on subsequent reheating after standing the associated water is expelled violently as steam, with consequent damage to the crucible and hazard to the operator.

The melting point of tin (232 °C) is low compared with that of the metals with which it is customarily alloyed, i.e. copper (1083 °C), antimony (630 °C) or even lead (327 °C). Nevertheless, alloying presents no problems if the procedures are correctly carried out, temperatures are accurately controlled and adequate time is allowed for dissolution of the added element. For certain purposes, especially when high proportions of an alloying element are needed, the use of an intermediate 'master alloy' is recommended.

From the melting aspect of tin alloys, the addition of an element such as copper, silver, antimony or lead alters the liquidus (according to the appropriate equilibrium phase diagram) and hence the melting temperature required. Of the above-named elements, antimony has the least effect on the melting point of tin, the addition of up to 10 per cent antimony raising the liquidus temperature by about 15 °C (see Fig. 41). On the other hand, copper and silver produce a marked increase in the liquidus temperature for quite small additions (see Figs. 25 and 42). Between the liquidus and solidus temperatures of a tin-rich alloy containing copper, for example, the compound $Cu_6Sn_5$ will be present in the melt as a dispersion of acicular crystals, causing the melt to be viscous and giving rise to difficulties during casting. It is therefore essential to hold the melt at a temperature high enough above the liquidus to ensure that the primary phase (such as $Cu_6Sn_5$) does not form before the metal enters the casting mould.

From the above it is clear that, if relatively high melting temperatures are to be avoided, the alloy content must be restricted for compound-forming elements such as copper, silver, nickel, iron, etc. In practical terms, 3 to 4 per cent copper is preferred in tin-base bearing alloys although some commercial materials do contain 7 per cent copper, while 5 per cent silver is the limit in tin—silver soft solders.

*Lead*
The addition of lead to tin reduces progressively the liquidus temperature of the melt until the tin—lead eutectic composition is reached at approximately 62 per cent tin, 38 per cent lead, melting at 183 °C. For lead contents higher than the eutectic, the liquidus rises again (see Fig. 39). The alloying of tin with lead to make up tin—lead solders presents no difficulties and for lead contents up to about 60 per cent the required quantity of lead as bar or sheet is added to the molten tin at a temperature of about 50 °C above the liquidus of the alloy being produced. With high-lead-content alloys, because the relative volume of liquid tin would be rather small compared with the bulk of solid lead to be

added, it is preferable to melt the lead first and to hold it at about 350 °C whilst adding the required quantity of tin, which melts and dissolves instantly.

### Cadmium
Cadmium is freely soluble in molten tin and presents no problems in making alloys except that of containing the possible hazards associated with the use of cadmium. In particular, overheating the alloy must be avoided.

### Bismuth
Bismuth also freely dissolves in molten tin and presents no difficulties.

### Antimony
In the preparation of pewter and certain solders and bearing alloys, the tin is melted and antimony added as solid pieces. Once it is wetted by the tin it dissolves readily. However, if it is not wetted quickly antimony will float and become oxidised, so that antimony powder is especially undesirable for making alloys. To ensure that wetting occurs, the antimony may be dipped in aqueous inorganic chloride flux just prior to addition to the molten tin, but great care should be taken to minimise spitting when the flux-covered metal touches the melt. In this way complete solution can be obtained without the bath temperature exceeding 400 °C. Traces of the flux will remain on the molten alloy surface so that metal for casting into ingots should be drawn from below the surface to avoid contamination and the surface of the melt should be skimmed. A frequently employed alternative method of adding antimony to a tin melt is to use a master alloy containing, for example, 50 per cent tin, 50 per cent antimony. Care should be taken to ensure that such an alloy, which has a fairly high melting point, is completely dissolved and that small pieces of alloying addition are not prevented from dissolving by becoming mixed up with dross (surface oxides) on the molten metal; the melt should be thoroughly stirred with a steel paddle. Where copper is also to be added (such as when making pewter or bearing metals) the master alloy may consist of a copper—antimony mixture which has been previously made up by melting copper and adding antimony, the product being analysed before use as a master alloy.

### Copper
Additions of copper are made to tin at temperatures of 300 to 400 °C, when clean copper wire or thin sheet rapidly dissolves. Pre-fluxing of the copper, for example in very dilute hydrochloric acid solution, assists with rapid wetting and hence dissolution. For some purposes tin-coated copper is preferable as no fluxing is then required. Copper may also be added as a copper—tin master alloy (for example, 50 per cent of each metal). The combined addition of copper and antimony is mentioned above.

*Zinc*
The presence of zinc has a pronounced effect on dross formation and even a fractional percentage of zinc increases the rate of oxidation dramatically. Because of this the surface of the melt should be flux covered when making up an alloy containing, say, 20 per cent zinc (cartridge lid metal). A fused inorganic chloride flux (for example zinc/ammonium chloride) is probably most suitable; this should be removed before casting the alloy. The temperature of the melt should be closely controlled to minimise oxidation which could result in the inclusion of hard particles of oxide in the casting.

*Silver*
Silver rapidly and easily dissolves in molten tin provided it is substantially free from sulphide tarnish films.

*Nickel*
Techniques similar to those described for copper are required for additions of nickel to tin.

## 7.2 CASTING

Tin and tin alloys can be cast successfully by a wide range of techniques including sand casting, gravity die-casting, pressure die-casting and centrifugal casting. The production of sound, dimensionally accurate, castings is aided by the absence of gas evolution and by low solidification shrinkage. An essential feature of the casting operation for tin and tin-rich alloys is to avoid entrapping in the casting the light oxide film which forms on the surface of the molten metal. Other general factors of importance include the need to reduce segregation to a minimum and to provide suitable arrangements for 'feeding' additional metal for castings with a long solidification range.

Typical procedures for different types of casting are given below.

### Ingots and slabs for rolling
Tin ingots, especially those for remelting, are often cast into normal pig or ingot moulds either individually or on a casting wheel or belt filled via a launder and molten metal pump. Slabs for subsequent rolling to sheet may be cast either horizontally in a shallow cast-iron mould or vertically in a book-type, water-cooled mould. The horizontal type of mould resembles a tray and for making pewter rolling slabs may typically be $800 \times 500 \times 50$ mm depth. In this type of open mould directional solidification is induced by water or air jets cooling the underside of the mould. This causes a depression due to shrinkage to be formed in the top surface of the ingot and this may be eliminated by 'feeding' (just before final solidification) a small amount of additional metal into the central contraction cavity. This operation may be repeated two or three times. To

prevent dross and oxide being cast with the metal, the surface of the tin alloy in the ladle must be skimmed before pouring and, preferably, the metal poured from below its free surface using a bridge across the pouring lip. Casting temperatures should be about 50 °C above the liquidus; higher temperatures should be avoided since they cause increased oxidation as well as coarse-grained ingots and may also increase segregation due to slower cooling of the casting. In high-antimony alloys (containing more than about 7 per cent antimony) primary intermetallic compound SbSn will be formed and, owing to its lower density, will tend to float on the surface. The presence of a small percentage of copper, which will precipitate first as acicular crystals of $Cu_6Sn_5$ dispersed in the melt, and the use of high rates of solidification, tend to prevent this.

When casting in vertical, book-type, metal moulds, which may be hollow and water-cooled, it is important to avoid splashing of the metal during the casting operation which would cause the formation of 'cold shuts' and hence laminations in sheet subsequently rolled from such ingots. A preferred procedure is to pour the metal into the mould through a vertical passageway formed by inserting a vertical strip in the mould, leaving a gap above the bottom of the mould of about 10 to 20 mm through which the molten metal enters and fills the main mould cavity without turbulence (Fig. 68). Alternatively the metal may be fed through a pipe or through a side section of the mould containing a slit or a series of holes. Tilting the mould almost to the horizontal position when starting to fill and then returning it progressively to the vertical is common practice when making pewter ingots or castings.

### Billets for extrusion

These billets are typically 100 mm diameter and 300 mm long and are cast vertically into steel 'chill' moulds using feeding to minimise the amount of piping due to contraction upon solidification. In modern automatic solder-billet casting machines about six thin-wall steel moulds (having a slight taper) are filled simultaneously via a molten-metal pump. After filling, water sprays are directed on to the exterior of the moulds and are progressively raised from the bottom to the top of the mould to give directional cooling from the base. Finally the moulds are automatically inverted and the cast billets fall into a conveyor system which transports them to be automatically machined and inserted in the extrusion press.

Billets of tin or tin-rich alloys for extrusion may also be formed by pouring the molten metal directly into the container of the press and then extruding after solidification. Difficulties can be encountered when using this simple process due to entrapment of internal oxide films, even though care may be taken to prevent the inclusion of oxide by controlling the casting process. To restrict access of oxygen, metal can be pumped from beneath the surface of a holding pot through a valve which opens after each stroke of the piston.

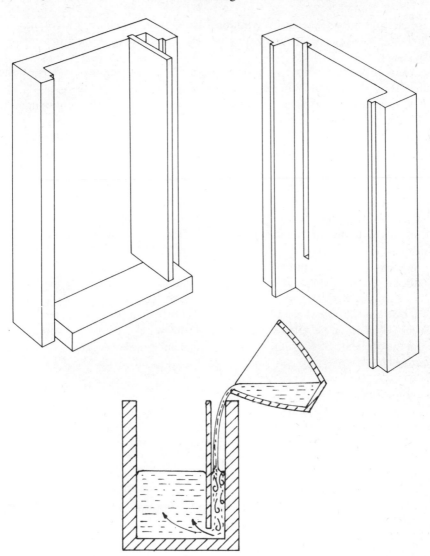

Fig. 68 – Diagram showing book-type mould for casting pewter slabs for subsequent rolling and principle of bottom filling.

## Continuous casting

In general the volume requirements for tin-rich alloys do not justify the use of
continuous casting processes. However, pewter (for subsequent rolling to sheet)
has been cast as wide slab on a standard horizontal casting machine using a
water-cooled graphite die. Experiments have also demonstrated the feasibility of

vertical continuous casting of cylindrical rods of solder alloys to be cut into billets for subsequent extrusion. In this case a water-cooled, oxidised-steel die was used.

A very old but unique form of semi-continuous casting is used for making thin sheets of tin—lead alloy for the subsequent manufacture of organ pipes. The metal is cast on to a table consisting of a cloth-covered stone slab approximately 4 m long and 1 m wide. A narrow wooden trough which has a gap beneath one vertical side, corresponding to the thickness of sheet required, is positioned across one end of the table and the trough runs smoothly along rails on each long edge of the casting table. The tin—lead alloy is allowed to cool in the ladle to just below its liquidus temperature before being rapidly poured into the trough, which is then moved by hand at considerable speed along the length of the table to leave behind a chill-cast sheet of alloy exhibiting on the under face the pattern of the cloth and on the upper face a eutectic cell structure. The latter gives rise to the name 'spotted' metal (Fig. 69).

A special form of continuous casting is used for soft solder stick. To produce a continuous length of thin strip of flat or triangular cross-section, the metal is poured at a controlled rate into grooves in the surface and around the periphery of a cast iron casting wheel which rotates in a horizontal plane. Peripheral speeds of about 0.3 m/s are employed. As it rotates, the wheel is cooled by water sprays from below and, after solidification, the solder is removed and cut into the required lengths.

### Sand-mould casting
Some pewter is cast into its final form in normal sand moulds with great facility. The cast surface is moderately rough and usually requires to be machined before the final buffing of the article to produce the traditional polished or satin finish.

### Die-casting
The low melting point and extreme fluidity of tin-rich alloys favour the production of sound castings of intricate design or pattern without difficulty and without damage to the mould. Antimony, copper and lead are the principal additions to tin for die-casting alloys and the alloys may be gravity die-cast or pressure die-cast. Various mould dressings are employed including plumbago, colloidal graphite, lime or chalk wash, etc., and once treated the mould can be used many times without further attention. The mould should be pre-heated to at least 100 °C to eliminate surface moisture which would cause defects in the castings as well as being a potential hazard to personnel.

For gravity casting the moulds are usually machined or die sunk in a low-phosphorus, flake-graphite, cast iron and the moulds are lightly dressed with talc to prevent wetting by the molten alloy. Before casting the moulds are pre-heated to a temperature sufficiently high to keep the metal liquid during pouring and then cooled with fine water sprays or by the local application of wet rags in such

Fig. 69 — Traditional method of casting 'spotted metal' (tin–lead alloy) sheet for manufacturing organ pipes. Considerable experience is needed to produce a sheet of uniform thickness. The 'spotted metal' surface of the chilled alloy can be seen in the foreground.

a manner that solidification is directional and shrinkage cavities and hot tears are prevented. The surface finish is adequate for polishing without intermediate machining and intricate engraved details in the mould surface are reproduced faithfully.

The traditional method of hand casting pewter is gravity die-casting in which the craftsman pours the molten alloy into a mould which is usually made of gunmetal or cast iron. The design of the mould with respect to runners, risers and gates does not always conform to modern foundry technology and many of the iron moulds in use today can be several hundred years old. Pewter tankards are often cast with only a local widening of the rim to form a 'runner' but the more difficult articles, such as large platters, are often cast in a vertical mould with a runner from which in-gates feed the mould cavity at the bottom, the centre and near the top in order to avoid cavities and cold shuts.

A typical tankard mould is seen in Fig. 70. The mould is preheated (often by immersion in the bath of molten pewter alloy), assembled, and the molten pewter is poured from a ladle into the aperture until it is full. Sometimes the mould is inclined from the vertical towards the ladle initially, and slowly tilted back to the vertical as it is filled. For certain articles such as tankards, localised quenching is achieved by the application of wet rags to a selected spot to eliminate casting contraction or other defects. For large platters, etc., the whole mould may be gradually immersed in a water tank to induce directional solidification from the lowest point in the mould.

Pressure die-casting is usually carried out in plunger-type machines which are operated either by hand or by hydraulic or pneumatic pressure (Fig. 71). In this type of machine the molten metal runs into a cylinder when the piston is withdrawn and is forced into the die aperture when the piston returns. Pressures around 2 to 3 $N/mm^2$ are customary, but higher pressures can be used provided suitable precautions are taken in preparation of working surfaces to prevent wetting and subsequent erosion by the molten alloy. Dies are often 'blued' to give an oxide film resistant to wetting by molten tin and then run in using lard oil as a dressing. After running in, no die lubricant is required.

### Slush casting

Slush casting is a method of die-casting confined to low-melting-point alloys. It is used for the production of small thin-wall hollow castings. In this method the molten alloy is poured into a hand-held mould and held for a short period to allow the metal in contact with the die wall to chill and solidify. The mould is then rapidly inverted and the still-molten excess metal is poured back into the melting pot. This technique is often used with pewter alloys for the production of hollow items such as handles for tankards, jugs, etc.

### Rubber-mould centrifugal casting

Large quantities of comparatively inexpensive costume jewellery and other

Fig. 70 – Five-piece cast iron chill mould for casting pewter tankard bodies.

Fig. 71 – Pressure die-casting of tin alloy (pewter) spoons.

decorative items are made by centrifugally casting tin alloys in rubber moulds. A synthetic rubber with Neoprene base will withstand about one thousand casts at 300 °C. To form the mould cavities, the master patterns of the articles to be cast are pressed between two circular discs of rubber which is in the partially cured condition. After the patterns have been removed, the rubber discs are vulcanised by heating. A central feeder hole is cut in the upper disc of rubber and from this channels are cut with a knife to each mould cavity in the lower disc.

When casting, the two halves of the mould are clamped together horizontally onto the rotating table of a centrifugal casting machine. The lid is shut so that the assembly is totally enclosed and the table is set spinning at high speed around the central axis. When the desired rotational speed is reached, a measured quantity of the molten alloy is poured into the central feeder, whence it is forced by the centrifugal action through the feed channels into the mould cavities (Fig. 72). Rotation speeds are usually up to 1500 r.p.m. applied for sufficient time to form and solidify the castings. After removing the castings from the mould, the sprues are cut off and the castings burnished. Jewellery produced in this way may subsequently be plated with precious metal or other decorative finishes.

## 7.3 FABRICATION

### Extrusion

For tin and tin alloys, both direct or indirect extrusion processes are used. Billets for extrusion (usually soft solder or pewter alloys) normally require one face to be machined flat before they are pre-heated and loaded into the extrusion press. This reduces the defects that would arise from the air gap between the remainder of the previously extruded billet and the end face of the newly inserted billet.

Solder wire may be extruded from the machined cast billet to produce the required final section shape and size, but smaller diameters (below 5 mm) are cold drawn by normal methods from the extruded product. Any segregation in the cast billet will manifest itself as a variation in composition along the length of the extruded wire, and care is required to maintain this variation within the required specification limits.

For the manufacture of flux-cored solder wire, extrusion is carried out at a fairly large diameter (for example 12 mm) using a bridge die having apertures through which flux is fed to the centre of the solder rod. In this manner single or multiple individual flux cores can be formed. Rosin-based fluxes are usually heated to 90 to 100 °C to make them fluid enough for feeding to this die; they may, in addition, contain a plasticising agent to maintain ductility of the rosin flux cores within the solder wire when it has cooled to room temperature.

When making cored solder wire, the original solder billet must be absolutely free from porosity, otherwise the wire is likely to break during drawing. Moreover the flux core must be continuous in order to ensure uniform soldering quality in

Fig. 72 — Rubber moulds for centrifugal casting of small figures, three at a time, in a tin alloy.

use; multiple flux cores were introduced to reduce the effect of random discontinuities within any one or two flux cores. Some manufacturers also extrude solder or pewter alloys as rectangular section rod for sale as melting stock.

*Impact extrusion*
A special type of extrusion process, impact extrusion, is used for collapsible tubes. These are nowadays mostly of aluminium but medical products are still packed in pure tin tubes. Collapsible tubes of tin are made by impact extrusion from 'slugs' which are specially shaped discs of pure tin normally containing 0.5 per cent copper as a hardener. The tin alloy is chill cast as a slab and rolled into strips from which the slugs are stamped. The slugs are simultaneously given a slightly conical shape to suit the die on the impact extrusion press. The ram of the press descends on the slug, ejecting the tin out of the die through a narrow annulus and forcing it into a tube shape enveloping the ram. The extruded tube is removed from the ram by compressed air fed through the ram. The neck is punched out and threaded to receive the cap. Finished tubes are printed and decorated, stoved, filled and the base sealed by folding and crimping.

**Rolling**
Tin and the majority of tin—lead alloys are easy to roll at room temperature. Tin—lead solder alloys may be rolled to thin sheet and cut into narrow ribbon; pure tin may be rolled down to foil as thin as 2 to 3 $\mu$m. In the case of pewter, many items are made by spinning or deep drawing sheet of 1 to 3 mm thickness. When starting from rolled sheet it is important that the mechanical properties of the pewter should be as uniform as possible. In particular there should be no directionality in the mechanical properties since anisotropic properties not only lead to non-uniform thickness of the walls of dish- or cup-shaped articles but also cause the appearance of regular protrusions or 'ears' around the rim, which are both inconvenient and wasteful of material. Moreover extreme anisotropy may lead to difficulties when forming complex shapes.

Broadly speaking, sheet pewter from which holloware is to be deep drawn or spun is produced from slabs cast originally about 50 mm thick and subsequently reduced by cold rolling to a final thickness, using a rolling sequence designed to eliminate anisotropy. The procedures which have been evolved include either cross-rolling, i.e. turning the sheet through 90° at a specific point in the rolling schedule, or alternatively heat treating the material for about 10 minutes at 175 °C at a certain stage preceding the final rolling reduction. With either technique the final reduction required to give minimum anisotropy is determined by the alloy composition and by the total reduction from the cast slab to the final thickness. The minimum level of anisotropy achievable decreases with the total alloying content. It is essential in the production of pewter sheet to use highly polished rolls having a bright finish, at least for the final reduction stage. All grit and dust must be eliminated to avoid their embedding in the soft surface.

**Spinning**

This process is applied in practice to the manufacture of certain items of pewter-ware, starting from circular discs stamped from rolled sheet pewter typically 1 to 2 mm thick. Generally the disc is clamped between the chuck and a wooden pattern or mandrel on a spinning lathe and force is applied to the rotating metal disc by forming tools made of wood or steel and lubricated with soap, tallow or olive oil. By making a number of passes of the tool, the metal is gradually forced to take on the configuration of the mandrel.

For some complex shapes the mandrels are made up from segments to facilitate removal from the finished shape. Some highly skilled craftsmen can spin 'on air', i.e. without using a solid mandrel. As the disc rotates the metal is first made to flow over the mandrel; this is known as 'drafting' (Figs. 73 and 74). From this basic shape, complex profiles and designs are then produced by deft manipulation of the various spinning tools.

During spinning, the wall thickness is reduced from that of the starting blank so this must be allowed for in the original material. Lathe speeds vary from about 800 r.p.m. for the larger discs to about 1200 r.p.m. for smaller

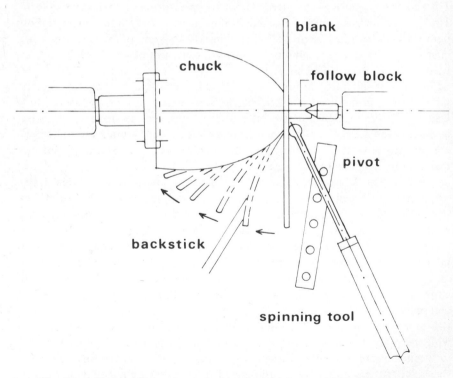

Fig. 73 – Diagram illustrating procedure for initial stages of 'drafting' a pewter vessel from a circular disc on a lathe.

Fig. 74 – Spinning of pewterware.

items. The use of a multi-speed lathe allows increased flexibility in dealing with a range of pewter thicknesses and different pewterware designs.

Some manufacturers produce tankards by first roll-forming pewter sheet into a cylinder, soldering this along the seam to make an open-ended tube known as a 'neck' and then spinning this to the final shape on a lathe. By this means flanges can be turned, contours modified when desired, incised lines added and unwanted metal trimmed away. A suitable base, also produced by spinning from a pewter disc, can then be attached by soldering. Automatic spinning machines are commercially available which utilise a tape-controlled series of forming tools to produce, for example, vases in pewter, with consequent savings in labour and time.

Articles having apertures smaller than their largest diameter (for example, vases) may be made from two separately spun sections soldered together, or alternatively, for small vessels, collapsible mandrels may be used.

### Pressing, deep drawing and stamping

Because of the high ductility of most tin-rich alloys, there is no problem in the production of pressings and stampings, such as medallions, using engraved steel dies, or in the deep drawing of cups. Drawn cups may be the starting point in the production of pewter holloware by subsequent spinning techniques. Art forms of thin pewter sheet may be formed on a wooden pattern by hammering or, if thin enough, by pressing into the mould recesses with a rounded non-metallic tool.

Certain shapes of solder pre-forms are manufactured by stamping them from rolled sheet or foil.

### Wire drawing

Pure tin is difficult to draw because it does not work-harden, but many solder alloys such as tin—lead, tin—antimony and tin—silver are commercially produced in a wide range of wire diameters.

The starting material is usually extruded rod of, say, 5 to 10 mm diameter and in the case of solder may incorporate several cores of flux. Normal wire drawing machines are employed using an oil—water emulsion as lubricant and relatively small reduction in diameter at each die. Final sizes may commonly be 0.5 to 2.5 mm diameter. Because of the inherently low strength of tin alloys, it is important that no casting defects such as porosity are present which will reduce the effective cross-section of the wire and lead to breakage during drawing.

### Machining

Because tin-rich alloys are soft and have high ductility, great care is required during machining to avoid distortion of the work and poor surface finish due to smearing. Small depths of cut with profuse lubrication are therefore necessary during turning on a lathe. The tool used should have an abnormally high rake on the top face to allow free movement of swarf, otherwise this will choke the cutting action.

Grinding of tin-rich alloys is not used although flat surfaces may be abraded with wet-and-dry silicon carbide papers provided the surface is flooded with kerosene or white spirit, as for example in metallographic preparation (see Appendix). Similar precautions have to be taken during drilling to avoid tin swarf welding on to and clogging the drill. Sawing may be carried out preferably using a coarse pitch saw (for example, a woodcutting saw) to ensure clearance of cuttings from the saw cut.

### Bronze casting

Tin bronzes and gunmetals (which contain copper, tin and zinc) are generally melted by similar procedures to other copper-rich alloys. Melting may be carried out in gas- or oil-fired crucibles or reverberatory furnaces, or in electric induction furnaces. The crucible material or the lining of the furnace is usually based on silicon carbide or carbon-bonded silicon carbide. The charge frequently consists of ingots of metal purchased to the required composition, although it is usually possible to include some selected recycled material such as runners and risers from previous castings. Bronze produced from scrap or secondary metal is melted, analysed and adjusted to the required composition by additions of virgin metals. When the metal is molten it is usual to put on the surface a cover flux consisting of borax and silica, but if the atmosphere of the furnace is neutral or oxidising in nature, a protective cover of charcoal may be used; this will minimise the solution in the metal of both oxygen and hydrogen, the latter being formed by combustion of gas or oil. When the atmosphere is oxidising in nature, as when air melting, oxygen may build up in the melt and will prevent pick-up of hydrogen gas; for this reason a slightly oxidising flame should ideally be used. However, under certain conditions in the gas- or oil-fired furnaces, there may be hydrogen derived from water vapour which dissolves in the molten metal, and this will give rise to porosity problems during casting. It is therefore usually necessary to degas the metal before casting. Degassing may be carried out by bubbling a stream of dry air or preferably nitrogen or carbon dioxide through the melt to displace dissolved hydrogen. If a tin bronze is being made by adding tin to molten copper, degassing is carried out before addition of tin. If an alloy of low oxygen content or of significant phosphorus content is required, then additions of copper—phosphorus hardener are made to the melt in order to produce, respectively, oxygen-free metal (about 0.015 per cent phosphorus) or a phosphor-bronze. Calcium boride can also be used as a deoxidant.

Zinc is added, perhaps as brass, to make gunmetal, although usually such alloys are obtained as ingots and remelted for casting, while lead may also be present to form leaded tin bronzes. Sometimes nickel is added to replace some of the tin in bronzes.

The temperature at which metal is poured into the mould after alloying depends on the section thickness being cast and the type of alloy. For most copper–tin alloys and gunmetals this will vary between 1150 °C and about 1250 °C. In the case of gunmetals, the temperature should be kept as low as possible and the melt maintained for the minimum time possible in order to reduce the quantity of zinc lost as zinc oxide fume. Furthermore these alloys require efficient fume extraction to avoid the health hazard associated with such fume. Losses of tin during melting and casting are usually minimal, but the lead content may be reduced because of vaporisation of lead at the elevated temperatures.

    Bronze and gunmetal are cast in sand or metal moulds, or may be continuously cast in special casting machines to produce rod or tubing suitable for further fabrication. In continuous casting the holding furnace containing the bronze is attached to a water-cooled graphite die of the required dimensions. The rod or tubing is withdrawn by an intermittently activated system of rolls so that the rod moves forward for a few seconds and then pauses for a further few seconds in order to induce solidification of a strong enough shell of metal to allow further movement forward of the rod (Fig. 75). The continuous casting

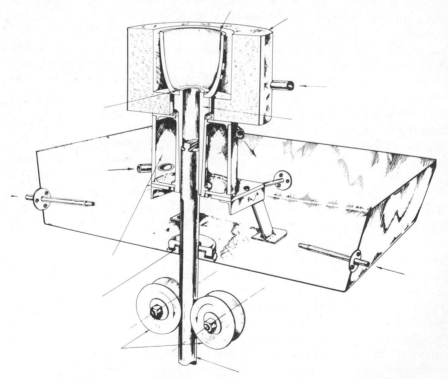

Fig. 75 – Diagram showing principle of semi-continuous vertical casting process
for bronze developed at the International Tin Research Institute.

process has almost entirely replaced the casting of bronze into large metal chill moulds. The latter produced ingots with relatively poor surface finish and significant piping due to contraction, so that a large proportion of the ingot had to be cropped before fabrication into rod or strip could be carried out. The continuously cast material may be taken directly to rolling extrusion, drawing, etc.

Sand moulds are mostly used where the finished casting is of intricate shape, such as for valve bodies.

Where permanent metal moulds are used for bronze casting they are usually made of grey cast iron with perhaps small alloying additions to improve the high temperature properties of the iron. Any necessary cores may be made of sand, plaster or graphite. A mould coating is required which allows escape of gases from the mould as the metal enters and also prevents reaction between the molten metal and any moisture retained in the mould which would give rise to surface porosity in the casting ('metal mould' reaction).

A related technique is the production of hollow tubing of large size by centrifugal casting. In this process the mould is usually made of steel, with a suitable mould dressing, and is rotated at the required rate while the molten bronze is poured into one end of the hollow mould through a suitable runner. At the completion of pouring, the exterior of the mould is sprayed with water to induce directional solidification. This process produces sound and dense castings which are often used for making cast bronze gear wheels and similar components. Diameters of 1 or 2 m are not unusual for such castings and the mould may be rotated at between 500 and 1000 r.p.m. to produce a pressure on the mould walls of about 0.3 $N/mm^2$.

**Bronze fabrication**

Alloys containing 10, 12 or 15 per cent tin are generally used in the as-cast condition, whereas alloys of lower tin content (up to around 8 per cent tin) are those intended for further fabrication. Cold rolling may be used to produce sheet in alloys containing 5 to 7 per cent tin, but hot rolling presents certain difficulties unless the tin content is restricted to 3 or 4 per cent tin. Continuously cast round bar containing up to 8 per cent tin may be drawn or may be first rolled and then drawn. Large diameter billets of cast round rod may be extruded at temperatures of around 700 °C, although this process is not normally used for bronze. Phosphor-bronzes are usually more difficult to hot or cold work because of the brittle phosphide phase present. For large reductions, as in the preparation of thin bronze sheet, intermediate annealing between the rolling reductions may be required; this annealing is usually carried out at above 350 °C in order to produce adequate softening.

Most of the difficulties of fabricating cast tin bronzes are due to the segregation that occurs during casting so that the brittle intermetallic δ phase is present. This affects both the ductility of the material and its hot strength and hence tends to induce surface cracking during hot working.

## FURTHER READING

1.  *Melting, Casting and Working of Tin and Tin-rich Alloys,* International Tin Research Institute Publication No. 456.
2.  *Working with Pewter,* International Tin Research Institute Publication No. 566.
3.  Hanson, D. and Pell-Walpole, W. T., *Chill-cast Tin Bronzes,* Edward Arnold, London, 1951.
4.  *Continuous Casting of Bronze,* International Tin Research Institute Publication No. 546.
5.  Ellwood, E. C., Prytherch, J. C. and Phelps, E. F., Experiments on the semi-continuous casting of bronze, *Journal of the Institute of Metals,* Vol. 84, p. 319, 1955.
6.  West, E. G., *Copper and its Alloys,* Ellis Horwood Ltd., (this series), 1982.
7.  Hedges, E. S., *Tin and its Alloys,* Edward Arnold, London, 1960.
8.  *Metals Handbook,* 8th edn., Vol. 5 Casting, American Society for Metals, Ohio, 1970.

# 8

# Tin and Tin Alloy Coatings

There can be little doubt that in applying tin to other metals the earliest purpose was for decoration. In early historical times all metals were valuable and the ability of tin to be spread thinly and evenly on the surface of other metals made it possible to achieve the full decorative effect of tin without using much material. Moreover, its low melting point made it readily usable by primitive methods. Although today it is customary to regard tin coatings for their protective properties rather than purely for decorative appeal, there is no evidence to suggest that the tinning of metals started that way. In fact the converse is probably true, since the earliest tinned objects generally had the coating on the visible outside surfaces rather than on the inside.

The properties of tin make it an ideal metal to use as a coating, both from the standpoint of applying the coating and that of its characteristics in service. Due to the low melting point of tin, and because it alloys readily with most other metals, tin coatings can be produced easily by hot coating methods such as immersing a suitably prepared object in a bath of molten tin, or by spraying molten tin on to a metal surface. A variation of hot-dipping – 'wiping' – in which tin is wiped over a fluxed and heated surface is used for local application, for example to one surface of a sheet or vessel. Another variation involves the application of tin powder to a fluxed and heated metal surface.

Tin coatings may also be produced by electroplating the metal from an aqueous solution of one of its inorganic salts (see Chapter 10). This is the method most commonly used for the production of tinplate for packaging and for alloy coatings.

The range of coating thicknesses practicable for the various processes are:,

| | |
|---|---|
| Electroless coatings, chemical replacement coatings | trace to 2.5 $\mu$m |
| Electrodeposits, for flow-melting | 0.4 to 7.5 $\mu$m |
| Electrodeposits, general | 2.5 to 75 $\mu$m |
| Hot-dipped coatings | 1.5 to 25 $\mu$m |
| Wiped coatings | 1.0 to 12 $\mu$m |
| Sprayed coatings | 75 to 350 $\mu$m |

In terms of tin consumption, undoubtedly the most important use of tin as a coating is for the manufacture of tinplate. This topic is dealt with in Chapter 9. The present chapter describes other tin and alloy coating applications.

The value and importance of tin as an industrial coating metal are enhanced because a wide range of tin alloys, each with its own specific properties and applications, may also be used to coat metal substrates, either by hot coating or electrodeposition. Among the most important tin alloys used as coatings are tin–lead, tin–copper, tin–nickel, tin–zinc, tin–cadmium and tin–cobalt (Fig. 76).

## 8.1 IMMERSION TINNING

Immersion plating, or electroless coating, usually involves a chemical replacement reaction between the substrate and the metal being plated. In the case of tin, true electroless plating by an autocatalytic process is demonstrably possible, but is of limited industrial application because of the inherently slow rates of deposition achieved to date. This means that only very thin coatings can be produced under conditions normally acceptable for industrial processes. Research is in progress, however, to develop electrolytes which will speed up the immersion tinning process.

The two main industrial applications of immersion tinning are both aimed at providing a thin tin coating to assist lubricity.

Sodium or potassium stannate are used for immersion tinning, in which a coating of tin is produced on a submerged object without the need for an applied current. In the automobile industry this process is used to produce a finish coating on aluminium alloy pistons and other parts to provide a non-scuffing surface during the running-in period. A coating of metallic tin about 2.5 $\mu$m thick is obtained after a short immersion.

In the manufacture of wire, stannous sulphate solution is used in 'liquor finishing', i.e. immersion tinning of steel wire with tin or with copper–tin alloy prior to drawing. This process is applied to both low- and high-carbon steel wire and the coating acts as a lubricant to ease the passage of the wire through the drawing dies.

Immersion tinning is sometimes used to coat tags and printed circuit boards in the electronics industry. For these copper-base alloys, acid solutions based on stannous chloride/thiourea are used. However, if prolonged storage is anticipated, immersion coatings are too thin to provide adequate retention of solderability.

## 8.2 ELECTROPLATED COATINGS

To electrodeposit tin coatings, alkaline sodium or potassium stannate, or acidic tin(II) sulphate or fluoroborate solutions may be employed. Alkaline electrolytes

Fig. 76 — The value of tin as a coating metal is enhanced by a wide range of tin-based alloy coatings, each with its own specific properties and applications. Illustrated here is a selection of articles electroplated with tin and tin alloy coatings including bright tin on a stein lid, tin–lead on a printed circuit, tin–zinc on hydraulic brake reservoir, tin–nickel on power connector and watch parts, tin–copper (speculum) on a dish, and red bronze on the medallions.

produce smooth but matte deposits on suitably prepared substrates and the plating solutions do not require the use of complex addition agents to produce smooth, coherent deposits. Acid electrolytes, on the other hand, require organic addition agents to produce good, fine-grained, coherent coatings. Because tin is plated from alkaline electrolytes in the quadrivalent form, the processes are less efficient in terms of power consumed for a given quantity of metal deposited. Nevertheless their independence of the need for addition agents makes alkaline plating processes easier to control.

Tin coatings plated from the most commonly used electrolytes are matte as plated, but relatively thin coatings up to about 8 $\mu$m may be brightened by a simple momentary fusion followed by quenching in water or another suitable medium. This process of 'flow-brightening' or 'flow-melting' is carried out for tinplate by conductive or inductive heating of the plated strip; for matte coatings on fabricated articles it is usually accomplished by immersion in a suitable heated oil bath. It should be noted that whilst the term 'flow' is used in describing these melting processes, the coatings are in general too thin to flow in the literal sense and there is no mass movement of the coating metal. However, because of the ease and rapidity with which tin forms alloys, 'flow-brightening' invariably causes an alloying reaction between the coating and the substrate; this assists in bonding the coating and hot-coated or flow-brightened plated coatings are especially adherent.

For some applications it is desirable that the coatings should be bright as plated and a number of electrolytes with specialised addition agents have been formulated for this purpose.

Sodium and potassium stannates form the basis of alkaline tin-plating electrolytes and were amongst the first to be used in electrolytic tinplate production. Electrolytes based on these stannates are very reliable and easy to operate and are capable of producing high-quality deposits. Advantages of stannate baths are that the solutions are not corrosive to steel and hence do not impose severe conditions on plating tanks, etc., and that no addition agents are needed. They also have high throwing power which makes them particularly suitable for plating fabricated articles. The process has some limitations — the bath must be operated hot (70 to 80 °C), maximum operating current densities (*ca* 270 A/m$^2$) are lower than can be used with acid tin-plating baths, hence deposition is slower and the range of anode current densities for satisfactory plating is restricted.

The use of potassium salts instead of sodium salts allows the cathodic current density to be raised appreciably without a loss in cathode efficiency and more concentrated solutions are possible because of the greater solubility of potassium stannate. As a result, faster plating rates are obtainable than with the sodium salt. Where advantage can be taken of the the higher plating speeds possible, the potassium stannate bath can usually justify its higher chemical cost, though to obtain maximum benefits special alloy anodes can be used. Insoluble (steel) anodes can also be used, in which case the tin is replenished by addition of tin salts. For efficient tin plating from alkaline electrolytes, the current density must be controlled between certain limits.

The principal acid electrolytes used in tin plating are based on stannous sulphate or stannous fluoroborate solutions. In the former process tin anodes are used and the bath contains stannous sulphate and free sulphuric acid and is stabilised against atmospheric oxidation by the action of phenolsulphonic or cresolsulphonic acids. The stannous sulphate-based bath can be operated at a high current efficiency (approaching 100 per cent in a static bath) but in order to plate a smooth, fine-grained, deposit certain organic addition agents such as gelatin and $\beta$-naphthol are required. Electrolytes based on this composition are in wide general use for batch plating of tin coatings. The tin is deposited as a matte finish but a bright appearance may be obtained by 'flow-melting'.

It is also possible to deposit bright tin coatings from an electrolyte based on tin(II) sulphate by the addition of certain brightening additives. There are a number of proprietary brightening additives in use today. Bright tin plating finds use in a range of applications such as finishing of electrical contacts, radio chassis, domestic articles and kitchen utensils. For some purposes bright tin plating is replacing cadmium plating or gold plating.

Tin fluoroborate solutions have good throwing and covering power provided suitable additives are present. Relatively high rates of deposition can be achieved since the high solubility of stannous fluoroborate enables solutions high in tin content to be produced. Some plating formulations have the disadvantage of sludge formation but this can be overcome by the use of certain organic addition agents. A principal use is in continuous plating units for the tinning of copper wire used for electrical conductors, in order to improve solderability and corrosion resistance. Tin fluoroborate solutions are also used for the barrel tin plating of components for subsequent soldering such as electrical contact lugs and for the backing of electrotypes. At least one large manufacturer of electrolytic tinplate uses a stannous fluoroborate electrolyte.

### Tin–lead alloy coatings

Fluoroborate electrolytes are used almost exclusively in tin–lead alloy plating and, according to the proportions of stannous and lead fluoroborates present in the bath, a whole range of tin–lead alloys can be electrodeposited. Free fluoroboric acid is present to prevent hydrolysis and to improve conductivity; the anode employed is usually of the same composition as the alloy to be plated. For smooth deposits addition agents such as peptone, bone glue or resorcinol are needed. Fluoroborate electrolytes operate at around 25 to 30 °C. A current density of 330 A/m$^2$ is used for still bath plating and up to double for continuous plating.

The growth of tin–lead plating has accelerated over the last few years, mainly due to the demands of the electronics industry. The coating is used to protect printed circuit conductive tracks (Fig. 77) and to plate other electronic

Fig. 77 — Automatic electroplating line for coating printed circuit boards with a thick layer of 60% Sn/40% Pb alloy.

equipment. It is also used in the manufacture of motor car radiators, in the printing industry to line electrotype shells, and as a coating for wire and strip.

### Tin–nickel alloy coatings

Tin–nickel alloy, which is an intermetallic compound of 65 per cent tin, 35 per cent nickel, is hard (around 700 HV) and remarkably tarnish- and corrosion-resistant. It is a whitish coating with a faintly pink tinge. The electrolyte is not self-levelling so that the brightness of the coating is similar to that of the substrate. This means that on a polished substrate a bright, lustrous deposit can be obtained.

A typical electrolyte for depositing tin–nickel contains stannous chloride dihydrate, nickel chloride and ammonium bifluoride, and plating takes place at 70 °C at pH 2 to 4. The cathode current density should be around 260 A/m$^2$ whilst the anode current density ranges from 100 to over 500 A/m$^2$ depending on whether alloy or separate tin and nickel anodes are used. Voltage is 6 V d.c. The deposit obtained with this electrolyte under normal conditions will be of constant composition, corresponding to the single phase intermetallic compound NiSn. This particular alloy can be produced only by electrodeposition. It is a metastable, non-equilibrium phase and will decompose on prolonged heating at 300 °C to the stable compounds $Ni_3Sn_2$ and $Ni_3Sn_4$.

A tin–nickel coating has good frictional and oil retention properties which make it a useful coating for certain watch parts and in pistons for brake mechanisms. Its tarnish resistance and good appearance make it an excellent finish for drawing and scientific instruments and for electrical equipment – especially for lighting fittings, reflectors and controls – as well as for decorative ware and for costume jewellery. Tin–nickel can be soldered and it is used in the production of printed circuit boards and in the electronics industry as a non-corroding finish for small contacts and components. It is frequently used as an undercoating for gold.

### Tin–zinc coatings

Tin–zinc coatings of several compositions can be plated from sodium or potassium stannate/zinc cyanide electrolytes, but the only coatings of commercial importance are those in the range 15 to 30 per cent zinc.

In its properties the coating behaves as if it were a mixture of tin and zinc rather than as an intermetallic compound as in the case of bronze or tin–nickel. In some marine environments the coating gives better corrosion protection than zinc, but in industrial environments it is inferior to galvanised steel.

The principal applications for tin–zinc coatings are for electrical and electronic equipment and in components for tools and mechanisms. Tin–zinc alloy is resistant to hydraulic fluids and so is used in some brake systems. It may also be used to coat fire extinguishers since it is less likely than zinc to produce loose corrosion products.

When fresh, or with careful storage, tin–zinc coatings are readily solderable; prolonged exposure to humid conditions reduces solderability, as in the case of zinc.

### Tin–cadmium coatings

Solutions comprising a mixture of sodium stannate with cadmium cyanide may be used as electrolytes for depositing a wide range of tin–cadmium alloys of differing compositions. Similar alloys may also be plated from fluoride/ fluosilicate solutions.

The behaviour of the tin–cadmium coatings is similar to that of tin–zinc. In practice, however, a coating containing at least 25 per cent cadmium, and preferably 50 per cent, is desirable for steel products in order to protect them from rusting at pores. As with tin–zinc, these coatings do not give outstanding protection to steel in ordinary outdoor exposure, but are capable of providing good protection in wet, marine environments.

### Tin–copper coatings

Tin–copper alloy coatings ranging in composition from 7 to 98 per cent tin can be obtained by suitable bath formulations. The most important commercial copper–tin alloys are the red and yellow bronzes ranging from 7 to 20 per cent tin and the white bronze alloy containing about 40 per cent tin which is known as 'Speculum'.

The electrolyte for bronze plating is a solution of sodium or potassium stannate with copper cyanide and a controlled excess of free alkali cyanide or alkali hydroxide.

Bronze electrodeposits containing from 10 to 20 per cent tin have been used as undercoats for nickel–chromium or tin–nickel deposits (q.v.). Tin bronze coatings have high wear- and corrosion-resistance and are used industrially for coatings (about 50 $\mu$m thick) on steel components of hydraulic equipment such as pit props which have to operate in wet conditions.

Speculum coating has a white, silver-like colour and is tarnish-resistant. It is suitable for domestic articles and because of its good resistance to sulphur compounds and other tarnishing agents in foods it is useful for tableware. However, the corrosion-resistant properties are very dependent on the correct alloy composition being achieved; the consequent close control of plating conditions required has hindered large-scale use of this attractive and useful coating.

### Brush plating

For electroplating tin or alloy coatings on selected areas, as for example in repairing coatings or for building up bearing alloys, a brush plating technique has been developed (Fig. 78).

It is an electroplating process that does not require a plating bath. A metal

Fig. 78 — 'Brush' plating, as a maintenance procedure, an interior bearing surface
with tin.

surface is built up from an electrolyte contained in an absorbent material such as cotton wool, which is wrapped around a hand-held electrode. The process has been designed so as to give excellent plating characteristics at speeds well in excess of those of a conventional bath system. The surface to be plated is made electrically cathodic and there is a range of portable inert anodes that can be used for the actual plating. The basic equipment consists of a special power pack, a selection of anodes and a range of specially formulated electrolytes.

Since plating occurs only where contact is made with the metal surface, the shape and size of anode is of prime importance in anode selection. There are some preparatory cleaning, deoxidising, etching and depassivating solutions, applied cathodically, which should be used in certain cases prior to electroplating. During the actual plating operation the speed of movement of the anode and the current density must be carefully controlled to ensure an even deposition of the metal and hence some skill is required.

The current density used in this process is of the order of 8500 A/m$^2$ which is relatively high and enables fairly rapid deposition of metal. By repeated applications any desired coating thickness can be built up. Within recent years the technique has been extended to include selective plating of tin alloys such as tin—lead, tin—nickel and tin—indium and this has further enhanced its usefulness in industry.

## 8.3 HOT COATINGS

The low melting point of tin and the ease with which it can wet and spread over a metallic surface make it an excellent coating metal to apply by hot-coating methods. Similar factors apply in the case of tin—lead alloys, in which it is the tin content which promotes wetting and alloying to the substrate. Whilst tin and tin—lead coatings can also be produced by electroplating, the converse is not true with other tin alloys; for example, tin—copper or tin—nickel coatings can be produced only as electrodeposits. In this section consideration is given only to articles coated after fabrication, not to pre-coated materials.

The widest application of hot-tinning is to provide a corrosion-resistant, hygienic coating and to give a pleasing appearance to corrodible metals (Fig. 79). Tin—lead alloy coatings are occasionally used as a cheaper alternative to tin where toxicity considerations do not apply. However, tin—lead alloys should not be used in contact with foods (see BS 3788 Tin Coated Finish for Culinary Utensils).

Another important purpose of hot-coating is to facilitate the soldering of steel, copper, brass and other metals. For such applications tin—lead alloy of near-eutectic composition (60 per cent tin) (Chapters 3 and 4) is often used because of its lower melting point (183 °C) and lower cost than tin. Alloys richer in lead may also be used for dip-coating of engineering assemblies such as heat exchanger cores and brass strip for fabricating radiator tubing.

Fig. 79 – A hot-tin coating is often used on welded steel pipes for aircraft refuelling
tanker lorries, to provide corrosion protection.

Hot-dipped coatings commonly cover a range of thicknesses from 5 $\mu$m to 15 $\mu$m, with uniformity of thickness. If, however, a somewhat uneven distribution of thickness can be tolerated, coatings as thick as 25 $\mu$m or more can be obtained. For these heavier coatings uniformity of distribution depends on the shape of the article and the skill employed in manipulation by the operator.

Hot-tinned coatings are usually produced by immersing the prepared work in a bath of molten tin. Articles which are too large to be dipped may be tinned by wiping. For this, the tin in stick, powder or molten form is applied to the prepared and pre-heated surface and wiped smooth. Hot-tinning by dipping can be restricted to prescribed areas by suitable masking.

An advantage of hot-tinning over electroplating is that bright, lustrous coatings are produced directly without the need for flow-brightening or polishing. Moreover, the fact that a smooth, continuous coating has been produced is evidence of good prior preparation and an assurance of good solderability.

The operation of hot-coating is more rapid than electroplating, normally requiring only a few seconds. It is usual practice to handle bulky articles separately or a few at a time using jigs. Very small parts are usually handled in bulk in baskets. Hot-dipping in tin–lead alloys is used to effect the simultaneous coating and soldering of automobile radiator blocks, heat exchangers and refrigerator cooling units (Fig. 80).

Hot coatings of tin or tin–lead can be applied easily to low alloy and carbon steels, to copper, brass and bronze. Cast iron can also be tinned readily with the aid of special preparative techniques. Other iron and copper-base alloys may be tinned provided that alloying elements which produce thick surface oxide layers, such as aluminium, chromium and silicon, are not present in substantial quantities. Nickel and most of its alloys are also readily hot-tinned. Lead and lead alloys may occasionally need to be pre-tinned as a preliminary to soldering; this may be done, but care and speed are necessary since lead dissolves rapidly in molten tin or tin–lead. Silver, if suitably prepared, and gold are readily tinned, but these metals, too, dissolve rapidly in tin, so care and skill are needed. Aluminium and its alloys are difficult to tin due to the tough surface films of alumina that readily form. However, by employing special techniques it is possible to tin aluminium; this is usually done as an aid to soldering rather than to provide a protective coating. Clean zinc dissolves very rapidly in molten tin and then oxidises to form a thick pasty dross on a tinning bath. The tinning of zinc articles is not therefore recommended if alternatives are available.

## Hot-dip coating techniques

The most important pre-requisite for successful hot-coating is correct surface preparation to ensure that a clean surface is available which the molten tin or tin–lead alloy can wet. Detailed procedures for tinning various metals are given in specialised handbooks. General principles only are outlined here.

Fig. 80 — Small heat exchanger units are assembled by dipping in a solder bath to ensure efficient thermal conduction during use between tubes and fins.

The operations involved in hot-dip coating are degreasing, acid pickling, fluxing, tinning, manipulation and finishing.

Degreasing involves the removal of all grease, oil and solid particles of dirt. It is normally carried out either by the use of organic solvents or their vapours or by treatment with aqueous alkaline solutions. Alkaline cleaning may be assisted by electrolytic action and solvent degreasing with the aid of ultrasonics. Rinsing following aqueous degreasing is essential.

Any of the common mineral acids are used for pickling to dissolve oxide films on the metal surfaces to be tinned. Choice of acid and pickling conditions depend on the substrate metal. Care must be taken in all pre-treatments to avoid entrapment of solutions which can lead to problems during tinning and possibly corrosion during service.

Fluxing immediately prior to hot-coating assists the molten tin or alloy to wet the metal surface. Fluxes are used in two ways, firstly as aqueous solutions in which the work is dipped before it is immersed in the molten metal and secondly, as a layer of fused salts on the top of the tinning bath. Both aqueous and fused fluxes are based on zinc chloride.

Dip coating involves lowering the prepared and fluxed articles through a fused flux cover into a bath of molten metal maintained at a temperature of 250 to 300 °C, leaving them immersed for a predetermined time and withdrawing. Withdrawal preferably takes place rapidly and where possible the flux cover should be drawn back so that the articles emerge through a clean surface. For high quality work, a two-pot tinning may be used. In this case the final bath has a layer of palm oil rather than fused flux.

Manipulation after dipping is required to obtain a coating of uniform thickness. Centrifuging inside a closed vessel is sometimes employed to remove excess· coating. Where a number of small items are tinned together, some form of mechanical separation is needed before the coating solidifies, to prevent 'soldering' together.

Finishing embraces a range of processes such as cooling, washing — important to remove all flux residues — and drying.


## 8.4 TERNEPLATE

Terneplate is normally defined as mild steel strip or sheet coated on both surfaces with a terne alloy. Terne alloy is generally understood to be a lead—tin alloy, or more correctly a series of lead—tin alloys, although some terneplate coatings use antimony to replace some or all of the tin. The tin content of terne coatings is usually within the range 8 to 25 per cent. The higher tin alloys (20 to 25 per cent tin) are often referred to as 'tin ternes' whilst the name 'ternecoated' is reserved for alloys containing 8 to 12 per cent tin, which constiture the majority of the production. Even the so-called 'lead coated' sheets, however, usually contain from 2 to 4 per cent tin as this helps to promote wetting and adhesion of the coating to the base. Pure lead, of course, does not wet steel and therefore cannot be used as a coating.

Terneplate is normally produced in thicknesses between 0.30 and 2 mm. The range of coating thicknesses is usually from 5 to over 12 $g/m^2$, as shown in Table 19. Of the coatings listed, the most commonly used are the four lighter coating grades. It should be noted that (unlike the system for tinplate) the coating mass values refer to the *total* coating on both surfaces. Normally this is equally distributed between the two faces of the sheet.

Terneplate is available in a variety of forming qualities (Table 20). There are generally three or four qualities corresponding to the formability of the steel base. 'Commercial' quality is intended for general fabricating purposes and is capable of withstanding bending and moderate forming or drawing processes.

### Table 19

Typical terneplate coatings

| Coating designation | Minimum coating weight (total both sides) ($g/m^2$) | |
| :---: | :---: | :---: |
| | Triple spot test | Single spot test |
| 001 | no minimum | no minimum |
| 050 | 50 | 40 |
| 075 | 75 | 60 |
| 100 | 100 | 75 |
| 120 | 120 | 90 |

### Table 20

Typical mechanical properties of terneplate

| Drawing quality Designation | Name | Maximum tensile strength ($N/mm^2$) | Minimum elongation (%) Gauge length 50 mm | 80 mm |
| :---: | :--- | :---: | :---: | :---: |
| 01 | Commercial | — | — | — |
| 02 | Drawing | 430 | 24 | 23 |
| 03 | Deep drawing | 410 | 26 | 25 |
| 04 | Deep drawing special killed | 410 | 29 | 28 |

For more severe drawing operations, up to three special grades, 'Drawing', 'Deep Drawing' and 'Special Killed' steel qualities may be specified.

The coating of terneplate adheres tenaciously to the base and will withstand any forming that the base steel will sustain. Indeed, the terne coating imparts a certain lubricity to the steel and can assist some drawing operations.

### Manufacture of terneplate

The low melting point of terne alloys and their ability, when molten, to wet and alloy with clean mild steel make coating by hot-dipping an ideal method of producing terneplate and the overwhelming majority is produced in this way. In recent years several proposals have been put forward for electroplating lead–tin coatings on to wide steel strip, but to date only one manufacturer is producing this material.

Terneplate is normally produced from cold-reduced mild-steel strip or sheet. Before being coated, the steel surface must be specially treated to ensure that it is perfectly clean and to promote wetting by the molten terne alloy. The treatment entails pickling in acid to remove oxides and immersion into the molten terne alloy through a layer of molten flux which floats on it at the entry end. The flux is usually based on zinc chloride or a mixture or solution of chlorides, often with hydrochloric acid added.

In the single-sheet process, the sheets are conveyed into and through the coating metal, held at a temperature of about 50 °C above the liquidus, via a series of submersed rollers. The leading edge of the sheet is guided to emerge upwards out of the coating pot at the exit end. At this point a layer of heavy oil covers the molten metal. In the 'single-sweep' method this oil, in conjunction with partially submersed coating rollers, controls the coating thickness.

For heavier quality coatings, and for better coating weight control, a double-sweep method is employed. In this the first coating pot, essentially as described above, is immediately followed by a second in which the sheet enters through a layer of molten oil rather than through a layer of chloride flux. Otherwise the details of the two coating pots are essentially similar. After coating the sheets are cooled on a conveyor, sometimes with forced air cooling.

The oil layer on the exit surface of the coating metal baths not only serves to prevent oxidation of the molten terne alloy but also leaves a residual oil coating on the sheet surface. Normally this is a heavier oil film than is required and so excess oil is removed by passing the sheets, after cooling, through a cleaning or 'branning' machine in which surplus oil is removed.

The various coating thicknesses may be obtained by suitable adjustment of the coating rollers and by controlling the angle at which the sheet emerges from the molten terne alloy.

Hot dip coating lines for the production of terneplate strip may be of the continuous or semi-continuous type (Fig. 81). In the latter each incoming coil is

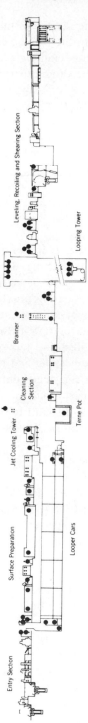

Fig. 81 – Diagram of modern continuous coating line.

processed as a separate entity, whereas in the fully continuous lines the leading edge of each new coil is welded to the trailing edge of the preceding one. This entails more complex and expensive handling equipment at both entry and exit ends of the line but leads to higher productivity. After coating the terneplate is either re-coiled or cut into sheets. A typical strip processing line for terneplate operates at a speed of 30 to 40 m/min.

The pre-treatment, coating and post-coating stages in strip coating lines reproduce those in the sheet process. Normally in-line pickling is practised and the entry of the strip into the flux-covered terne alloy may be preceded by a pass through a tank containing flux solution. On emerging from the coating pot, through the oil layer, the strip travels vertically until the coating has solidified. This entails a high vertical tower section in the modern higher speed lines, although cooling may be assisted by air blasts. Post-coating treatments involve removal of surplus oil, as in sheet coating.

As with all coating processes, it is important to obtain a perfectly clean surface on the basis metal prior to coating. This is especially true of hot dip coating with the lower tin content terne alloys, for it is the tin which provides the essential wetting and alloying. Coating problems can arise if the steel is not correctly prepared. The usual preparation treatment is by pickling in hydrochloric acid, with or without electrolytic pickling. Unlike the light gauge steel for the production of tinplate, steel for terneplate is usually passed to the annealing department without pre-cleaning and hence carries a residual oil film. If this is excessive, carbonisation can occur during annealing, even though the heating and cooling cycles are performed in a controlled inert or reducing atmosphere. The carbon film can be difficult to remove by conventional pickling and specialised treatments are sometimes adopted.

### Properties and applications of terneplate
The properties imparted to steel by a terne coating are corrosion resistance, solderability and improved drawability due to the lubricity afforded by the lead—tin alloy. These properties have led to specialised uses for terneplate.

One of the most important properties of the terne alloy coating is its resistance to corrosion by petroleum products. This property provides its largest use which is as a material for fuel tanks and filler pipes in cars and lorries. The coating is extremely adherent and does not flake during the rigours of fabrication or if the steel suffers mechanical damage. This eliminates the risk both of foreign particles which could clog fuel lines and bare spots on the exterior of the tanks which would invite rusting.

Additonally, the excellent formability of terne makes possible the drawing of the often complex shapes of the modern fuel tank (Fig. 82). These are frequently made from two pressings which can readily be welded together, while filler pipes and fuel lines are easily and quickly attached by soldering. The automobile industry also uses terneplate for petrol, oil and air filters, radiator

Fig. 82 – Small fuel tank for tractor made in two pieces from terneplate by deep-drawing. Terneplate offers a combination of the mechanical strength of the steel substrate with resistance to corrosion both on the internal and external surfaces.

header tanks, radiator brackets, car heaters, heat exchangers and water expansion tanks in lorries.

The electrical industry recognises the corrosion protection and good solderability of terne-coated steel in such products as conduit and junction boxes and in the use of light-gauge terne strip for flexible cable sheathing, as well as for outdoor lighting fitments. In the allied electronics industry, manufacturers of radio and television equipment use terneplate for a number of parts including chassis and cases.

In the USA, ternplate is used for a wider variety of applications than in Europe and, in particular, it is frequently used as a building material. The resistance of terneplate to atmospheric corrosion is remarkably demonstrated by its use as a roofing material and there are numerous instances of buildings with such roofs that have survived a century or more. In the USA the roof of the historic home of General Andrew Jackson in Tennessee is still tiled with the original terneplate used when it was rebuilt in 1835. As an example of longevity in an outdoor exposure test this must rank among the leaders!

In addition to its natural corrosion resistance, terneplate can be readily painted without any pre-treatment other than cleaning.

Other building uses include rainwater goods, fire-resisting doors, ducting, outdoor signs (Fig. 83) and components for roller blinds. Among the applications of terneplate to be found in office and industrial premises are oil heating furnaces and equipment, office and laboratory furniture, deed boxes and heavy-duty cleaning equipment.

## 8.5 OTHER COATING METHODS

### Wipe coating
A special case of hot coating sometimes adopted for tin and tin alloys is wipe coating. The process is used when it is desired to apply the coating to a localised area of an article. The most usual application is for wipe tinning of cooking vessels, to apply or renew a tin lining on the inside of the copper cooking utensils used in the catering industry.

Preparation of the surface prior to coating is carried out as described for hot tinning. The article to be tinned is then heated to above the melting point of tin and the surface is fluxed by sprinkling powdered ammonium chloride on the surface. Molten tin together with flux is poured on to the prepared surface and spread quickly over the entire area with cork or hemp wiping pads.

Wipe tinning is a skilled job if a smooth and uniform coating is to be obtained. Temperatures are judged largely by inspection and experience and the work must be carried out rapidly to avoid oxidation and undue formation of intermetallic compounds between the coating and the basis metal.

Solder is sometimes used as a filler in sheet metal work, especially in the automobile industry to smooth out joins and imperfections. Wipe coating of the

Fig. 83 – Large letters for industrial signs made up from terneplate components.
They are usually subsequently painted.

area is frequently a preliminary to solder filling. The solder is itself applied by a wiping technique using solder bars melted by blow-lamps. The filler is smoothed with wooden tools.

### Roller coating

Roller coating is another variation of hot coating. It is employed in the production of printed circuit boards. As described elsewhere, it is desirable for the copper circuit tracks to be coated with tin or solder alloy to preserve solderability.

In a roller coating machine the prepared and fluxed printed circuit board is passed between a pair of driven rollers. The lower one is a plain or grooved metal roller which rotates half-submersed in a bath of molten solder. This roller picks up solder from the bath and transfers it to the circuit board. The solder bath carries a flux cover to prevent drossing. Coating thickness is controlled by the contour of the coating roller and the pressure and nip of the rollers.

### Spray coating

Tin-base alloys, especially fusible alloys (Chapter 4), because of their low melting temperatures, are readily adaptable to spray coating. A typical application for spray coated fusible alloys is as a metallised lining for non-metallic (usually plastics) moulds for low temperature casting. The technique very accurately reproduces fine detail.

The equipment for spray coating is normally a hand gun supplied with electric power for heating and compressed air for spraying. The fusible alloy is fed to the gun in stick form. This is melted by electric heating and in the alloy melting chamber is mixed with a supply of heated compressed air before being ejected via an adjustable spray nozzle on to the work. Usual coating thicknesses by this method are less than 1 mm thick, but spray coatings can be built up by successive applications to several millimetres thick.

### 'Quench' coating — the H63 process

A unique method of applying tin and solder coatings has been devised by Friedrich Heck and designated the H63 process. It is used for coating small parts for the electrical industry.

Copper, copper alloys or steel can be hot tinned satisfactorily if the parts are preheated and then quenched in an aqueous solution of zinc chloride flux in which a finely divided metal powder is dispersed. The preheating temperature required is 280 °C for pure tin or about 220 °C for the 63 per cent tin—37 per cent lead eutectic solder alloy. The resulting coating has a matte appearance and flow-melting is required in order to obtain a uniform, bright coating.

Parts to be coated are carried at a uniform speed through a conveyor-belt furnace where they are heated to the required temperature. From the furnace, the parts slide down a ramp into a rotating drum containing the flux/metal dispersion. The drum contains baffles which keep the liquid agitated and the metal powder evenly dispersed. The baffles also serve to eject the hot-tinned parts at the end of the cycle and these are then rinsed in water.

The coating thickness is controlled by a number of factors including the furnace temperature, the wall thickness of the parts, thermal conductivity, the concentration of the metal powder and the temperature of the flux/metal dispersion. To increase any or all of these produces a thicker coating. By suitably varying the conditions, coatings of a predetermined optimum thickness may be obtained. Coating thicknesses may range up to 20 $\mu$m or even greater. In order to obtain the minimum heat capacity, minimum wall thicknesses for copper and copper alloys of about 0.5 to 1.5 mm and for iron of about 1.0 to 1.5 mm are essential for hot-tinning with the eutectic tin–lead alloy.

One interesting feature of this process is that alloy coatings may be produced by using a mixture of tin and lead powders rather than a pre-alloyed solder powder. In this case the composition of the coating depends on the proportions of the mixture. This can reduce the inventory of the alloy powders required for a range of coatings.

In this tinning process it is possible to tin in succession a variety of metals – copper, copper alloys, steel, etc. – without replacing the flux. Unlike conventional hot-tinning, in which certain constituents are dissolved in the bath, an advantage of the H63 process is that no metal is dissolved.

### Mechanical plating

Mechanical plating, sometimes called 'peen plating', is a coating method which dispenses with the need for either heat or electricity. It can be used for coating small hardware parts of complex shape with tin or certain tin alloys.

The process combines the techniques of barrel finishing and cold welding. Plating is performed in a modified tumbling barrel using the required coating metal in powder form and containing also an impacting material. In the modern process the impacting medium is usually small-diameter glass spheres, although other materials have been used.

Prior to coating, the surfaces must be completely clean. Preparation includes degreasing by alkali or solvent cleaning and acid pickling. Frequently the surfaces are given a very thin copper coating by short-time immersion in hot copper-sulphate solution.

The prepared articles are placed in a tumbling barrel together with a controlled quantity of tin or alloy powder and water containing a proprietary chemical 'promoter' which assists adhesion of the coating. Tumbling at room temperature continues for a time between 20 min. and 1½ h, depending on the items being coated. Coating thicknesses achieved by this method are around 15 $\mu$m.

Whilst tin can be plated by this method it is more commonly used for tin alloys, especially for tin–cadmium and tin–zinc coatings.

## FURTHER READING

1. Price, J. W. *Tin and Tin Alloy Plating,* Electrochemical Publications, Ayr, 1983.
2. *Modern Electroplating,* ed. Lowenheim, Wiley, New York, 1974.
3. Mechanical plating with tin–cadmium alloy, *Tin and its Uses,* No. 112, p. 4, 1977.
4. *Practical Hot-Tinning,* International Tin Research Institute Publication No. 575.
5. MacKay, C. A. and Barry, B. T. K., Terneplate: production, properties and applications, *Proceedings of the 5th International Lead Conference,* Paris, 1974. (International Tin Research Institute Publication No. 502.)
6. *Instructions for Electrodepositing Tin,* International Tin Research Institute Publication No. 92.
7. *Electroplated Tin–Nickel Alloy,* International Tin Research Institute Publication No. 235.
8. *Electrodeposition of Tin–Lead Alloys,* International Tin Research Institute Publication No. 325.
9. *Tin versus Corrosion,* International Tin Research Institute Publication No. 510.
10. Heck, F., H63: a new method of hot-tinning, *Tin and its Uses,* No. 81, p. 14, 1969.

# 9

# Tinplate

---

Tinplate is by far the most important industrial outlet for tin since it accounts for around 40 per cent of the entire world consumption of tin with average usage of around 80 000 tonnes of tin per annum. Therefore it is justifiable in a monograph on tin that tinplate should occupy a separate section.

## 9.1 HISTORICAL DEVELOPMENT

The tinplate industry was probably the earliest to employ pre-coated sheet for the production of finished articles rather than to coat metal objects after their fabrication. The idea of tinning hammered iron sheet, so producing a durable material from which objects could afterwards be fashioned, appears to have originated in Bavaria in the fourteenth century. The practice of making tinplates spread through Europe in the fifteenth and sixteenth centuries and appears to have reached Britain early in the seventeenth century, although the industry was slow to develop until the end of that century.

Whilst the history of tinplate is one of continuous technical innovation and improvement right up to and including the present day, both in ways of making and using the product, there have been a number of major innovations which have marked significant milestones in the long history.

The first of these was the introduction of rolled iron sheet instead of the earlier hammered material. The exact date is not known, but certainly by 1730 John Hanbury of Pontypool, generally regarded as the founder of the modern tinplate industry, was producing tinplate by hot-dipping rolled iron sheets. The early use of tinplate was for the fabrication of household utensils — pots, pans, plates, drinking vessels, boxes, candle-holders, lanterns and the like — and many examples of early tinplate domestic goods are to be found in the museums. However, the most important stimulus to the growth and importance of the industry was the invention of a means of preserving sterilised foods in sealed containers by the Frenchman, Nicolas Appert, in 1810 and the adaptation of his process to the preservation of food in tinplate canisters by John Hall and

Bryan Donkin, who in 1812 produced the world's first 'tinned' or canned foods at their factory in London. From these small beginnings in the early decades of the nineteenth century the growth of the tinplate industry has been linked with that of packaging and today around 90 per cent of the 12 to 13 million annual tonnes of tinplate are used for packaging in one form or another.

Amongst the other major milestones in the industry have been the introduction of rolled steel for rolled iron in the latter half of the nineteenth century and the adoption of continuous strip rolling processes for the production of the steel base. These were first introduced in 1927 but it was not until after the Second World War that virtually all the hand sheet mills were closed down. Within the past 30 years the most important innovations in the steel manufacturing and processing methods have been the adoption of oxygen steel-making techniques, and more recently of continuous casting, in the development of continuous strip annealing, and in the advancement of steel rolling methods. On the tinning side undoubtedly the most significant advance was the introduction of continuous strip electroplating. This began in Germany in 1934, was adapted and developed in the USA in the 1940s and has today virtually ousted all hot-dip production to the extent that some 98 per cent of the world output today is 'electrolytic tinplate', made by the continuous electrodeposition of tin on to continuously cold-rolled and annealed steel strip.

### Description and nomenclature

Tinplate is defined in national and international specifications as thin low-carbon or mild steel sheet or strip coated on both surfaces with a thin layer of tin. In actual fact this simple definition obscures rather than reveals the fact that the term 'tinplate' is in practice a generic name for a wide range of products having differing mechanical and corrosion-resistant properties and capable of being used for a large diversity of applications.

By conventional coating terms, the tin coating of tinplate is extremely thin. In common with usual coating nomenclature, the coating on tinplate is usually expressed nowadays in terms of weight (or, more strictly, mass) per unit area. The common range of tin coating masses expressed as grammes per square metre of surface is from 2.8 to 15 $g/m^2$ per surface for electrolytic tinplate and from 11 to 22 $g/m^2$ for the less common hot-dipped tinplate. The usual range of steel thicknesses employed for tinplate is from 0.15 to 0.5 mm.

When tinplate is manufactured by the electrolytic coating process it is possible, deliberately, to apply a different, controlled tin thickness to each of the two surfaces. This makes it possible to provide the correct degree of corrosion protection for the differing environments of the inside and outside of a container. Tinplate having such differing coatings is known as differential electrolytic tinplate, or, more simply, differential tinplate. During its manufacture, such tinplate has its characteristic coatings indicated by an internationally accepted code of surface marking which can be observed as a line pattern, often on the inside of a can. It is because the possibility exists to produce unequal coatings on both surfaces that the coating thickness is nowadays quoted for each surface separately and the appropriate numbers are prefixed with the letter 'E' (equally coated) or 'D' (indicating differentially coated). Thus a quality indicated as E11.2/11.2 would have 11.2 $g/m^2$ on each face whilst the quality D11.2/2.8 would have 11.2 $g/m^2$ on one surface and 2.8 $g/m^2$ on the other.

Table 21 indicates the usual range of coatings as specified in international and national standards, together with the corresponding thickness of coating in linear terms.

**Table 21**

Electrolytic tinplate coatings

| Designation | | Nominal coating mass | Nominal coating thickness |
|---|---|---|---|
| ISO | ASTM | $g/m^2$ per surface | $(mm \times 10^{-4})$ (each surface) |
| E2.8/2.8 | 25 | 2.8 + 2.8 | 3.8 + 3.8 |
| E5.6/5.6 | 50 | 5.6 + 5.6 | 7.7 + 7.7 |
| E8.4/8.4 | 75 | 8.4 + 8.4 | 11.5 + 11.5 |
| E11.2/11.2 | 100 | 11.2 + 11.2 | 15.4 + 15.4 |
| D5.6/2.8 | 50/25 | 5.6 + 2.8 | 7.7 + 3.8 |
| D8.4/2.8 | 75/25 | 8.4 + 2.8 | 11.5 + 3.8 |
| D11.2/2.8 | 100/25 | 11.2 + 2.8 | 15.4 + 3.8 |
| D11.2/5.6 | 100/50 | 11.2 + 5.6 | 15.4 + 7.7 |
| D15.1/2.8 | 135/25 | 15.1 + 2.8 | 20.7 + 3.8 |

Whilst on the subject of nomenclature it is of interest to point out that tinplate is normally bought and sold on an area basis rather than on a tonnage basis, as are most steel products. The reason is, of course, that the area determines the number of cans or components that can be produced from a consignment. In UK and Europe the usual unit of commerce is the SITA (for System International Tinplate Area) of 100 $m^2$.

There are other, mainly older, systems of nomenclature for describing tinplate, some of which still persist in certain countries. In general, however, there is international agreement to move towards the system described above.

## 9.2 PRODUCTION OF TINPLATE

As will be realised from the foregoing, tinplate is basically a steel product, since the coating on average represents only around 0.6 per cent by weight of the finished product. Thus a description of the manufacturing processes inevitably includes a description of steel processing.

Figure 84 is a flowsheet indicating the production processes for tinplate. It will be seen that there are several alternative routes, which are chosen according to the properties required of the finished product.

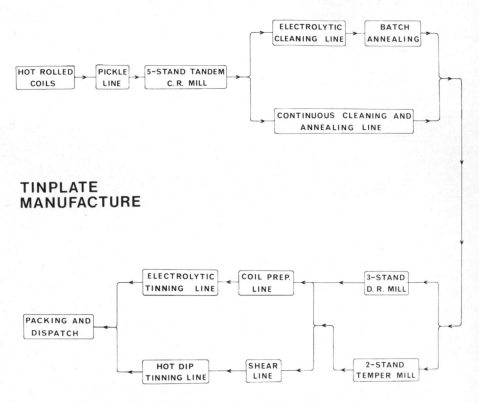

**TINPLATE MANUFACTURE**

Fig. 84 — Schematic flow-sheet of the tinplate manufacturing process.

The steel used for tinplate is a low-carbon mild steel usually containing less than 0.1 per cent carbon. The composition of the other elements is closely controlled as these govern to some extent the mechanical and corrosion characteristics of the steel (Table 22).

Most tinplate steels are produced by oxygen steel-making processes. At present the majority of plants cast the steel into ingots of up to 30 tonnes, although continuous strand casting is being introduced. In steel produced by the ingot route, the ingots are first heated uniformly and then rolled in powerful reversing slabbing mills to slabs 100 to 250 mm thick and of a width approximating to that of the finished strip, usually in the range 700 to 1000 mm.

**Table 22**

Typical compositions of steels for tinplate manufacture

| Element | Cast composition, maximum | | |
|---|---|---|---|
| | Type D | Type L | Type MR |
| Carbon | 0.12 | 0.13 | 0.13 |
| Manganese | 0.60 | 0.60 | 0.60 |
| Phosphorus | 0.020 | 0.015 | 0.020 |
| Sulphur | 0.05 | 0.05 | 0.05 |
| Silicon | 0.020 | 0.010 | 0.010 |
| Copper | 0.20 | 0.06 | 0.20 |
| Nickel | – | 0.04 | – |
| Chromium | – | 0.06 | – |
| Molybdenum | – | 0.05 | – |
| Other residual elements, each | – | 0.02 | – |

Type D steel — aluminium killed, sometimes required to minimise severe fluting and stretcher-strain hazards or for severe drawing applications.

Type L steel — low in metalloids and residual elements, sometimes used for improved internal corrosion resistance for certain food product containers.

Type MR steel — similar in metalloid content to Type L but less restrictive in residual elements, commonly used for most tin mill products.

Slabs produced in this way or by flame cutting from the continuously cast steel, are reheated to around 1300 °C and hot-rolled through a series of single pass mill stands, up to 10 in all, of which the final six stands are usually in the form of a continuous tandem mill. This sequence of operations reduces the strip gauge to between 1 and 3 mm depending on the desired end product. The exit speed at the final roll stand may be as high as 1200 m/min, after which the strip is subjected to rapid controlled cooling by water sprays and coiled whilst still hot, around 650 °C.

Rolling the steel whilst hot causes oxide scale to form on the surface and this must be removed before the next stage of rolling. Scale is removed by pickling the steel in hydrochloric or sulphuric acid in a pickle line which is essentially a series of shallow tanks, each about 25 to 30 m long, through which the strip is passed. After pickling to remove all surface oxides, the strip is rinsed, dried, coated on both surfaces with a heavy lubricant to minimise oxidation and facilitate cold-rolling, and finally re-coiled.

The principal cold-rolling operation, which reduces the steel thickness to the desired finished gauge (except for double-reduced tinplate (*q.v.*)), is usually undertaken in a five- or six-stand tandem cold-rolling mill. This imparts a controlled reduction of around 80 to 90 per cent. Modern mills operate at speeds of up to 2000 m/min and are equipped with sophisticated control mechanisms for gauge and shape control.

After removing all traces of rolling lubricant from the steel strip by treatment with alkaline solutions in a cleaning line or section, it is annealed by heating in a non-oxidising atmosphere which relieves the stresses imparted by the severe cold reduction and restores a measure of ductility to the steel. Steels for tinplate are annealed at temperatures up to 680 °C, i.e. below the $\alpha$—$\gamma$ transformation temperature. The purpose of this 'process annealing' is thus to achieve recrystallisation and grain growth rather than to induce phase changes.

As indicated in Fig. 84, there are two alternative annealing methods, namely batch and continuous annealing. In the former, coils of steel are stacked with axes vertical, and sealed into steel covers into which a non-oxidising gas is injected. The stacks of coils are heated and cooled through a controlled thermal cycle which may last 50 h or longer. In continuous annealing, the steel strip is uncoiled and travels as a single strand at up to 450 m/min through a tower furnace with a controlled thermal cycle. The properties of the strip which result from these two processes differ slightly on account of the differing crystal size and orientation resulting from the different thermal cycles.

After annealing, as again indicated in Fig. 84, there are two alternative processing routes. Generally the steel strip receives a further light cold reduction of from 1 to 4 per cent in a two- or three-stand skin-pass or 'temper' mill. This temper rolling takes place without strip lubrication; its purposes are to produce the formability appropriate to the intended end use and to impart the desired surface finish, by the use of suitably prepared work rolls.

For certain applications a harder and stiffer tinplate may be suitable. This can be produced by substituting for temper rolling a second cold reduction of 30 to 40 per cent in a two- or three-stand mill. This heavier reduction requires the application of rolling lubricants. Material produced by this process is called 'double-reduced' tinplate. It is a stiff, strong product with very directional mechanical properties.

The steel may be coated with tin either by immersing it briefly in molten tin (to produce 'hot-dipped' tinplate) or more usually by electrodeposition from an aqueous solution. Prior to being tinned, the steel is inspected for defects such as pinholes, surface defects or off-gauge material, and these are cut from the strip and rejected. Generally the mill edge is removed and the strip trimmed to finished width. For hot-dipped timplate, the strip is cut to sheet sizes which are

Fig. 85 – Entry end of a 100 000 t.p.a. electrolytic tinplate line. (Courtesy: Rasselstein A. G.)

then tinned individually. For electrolytic tinplate production the strip is built up
into coils of optimum size by welding strips together.

Worldwide, over 98 per cent of tinplate is produced by the electrolytic
process and hot-dipping has been entirely superseded in the major producing
countries of USA, Western Europe and Japan.

Electrolytic tinplate is produced by the continuous electrodeposition of tin
directly on to steel strip moving at high speed through a processing line (Fig. 85).
The entry section of an electrolytic tinplate line comprises uncoilers, a shear to
trim the ends of the strip square, a welder to join the leading end of one strip
to the trailing end of the preceding coil and a strip accumulator. This latter is
either a series of deep pits into which the strip hangs or, in most modern lines,
a looping tower in which the strip passes over two sets of rollers, the lower of
which is capable of being raised and lowered. The purpose of the strip accumulator is to build up a reserve of strip to be drawn upon when changing coils, so
that the electroplating section is able to run continuously at a steady speed and
hence make coating thickness control easier. Most entry storage looping devices
permit about 1 min for coil changing.

In the preparation section of the line, the strip passes successively through
tanks containing alkaline solutions to remove oils and grease, rinse water, and
acid to remove residual oxide. The alkaline cleaning and acid pickling processes
are usually electrolytic. After a final thorough rinse, the strip passes through a
series of tanks containing the tin plating electrolyte. During its passage through
the plating tanks the steel strip is in contact with conducting rollers so that the
strip forms the cathode in the plating cell. The plating tanks have tin anodes
which progressively dissolve as the tin is plated on to the strip. Spent anodes are
removed and replaced periodically. In a typical line, plating currents of up to
150 000 A at a potential of 20 V d.c. are employed.

After plating, the tin coating has a dull, matte, white appearance and for
most purposes the strip is 'flow-brightened' in order to give the coating its
familiar bright, shiny appearance and to improve performance. Flow-brightening
in a tinplate line is achieved by heating the strip momentarily by conductive or
inductive heating to slightly above the melting point of tin and then plunging the
strip into water. This process causes reaction at the molten tin/steel interface to
form a thin layer of intermetallic compound ($FeSn_2$) (Fig. 86) which plays an
important part in the corrosion behaviour of the material when used as food
cans.

For the manufacture of tinplate the two electrolytes in most common use
are based on stannous sulphate with phenolsulphonic acid or on stannous
chloride/fluoride, commonly known as the 'Ferrostan' and 'Halogen' electrolytes respectively. There is also a limited use of acid stannous fluoroborate and
alkaline stannate solutions. All except the alkaline solutions require the use of
organic addition agents to obtain smooth, coherent deposits. The electrolytes
used for tinplate manufacture are of patented compositions.

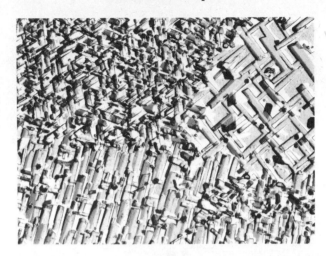

Fig. 86 – Transmission electron micrograph of the layer of FeSn$_2$ in electrolytic tinplate formed during the flow-melting operation. The overlying tin coating has been selectively dissolved prior to making the replica. (Magnification ×8000.)

In the finishing section of the electrolytic tinplate line, following flow-brightening, the strip is given a chemical or electrochemical 'passivation' treatment in an aqueous sodium dichromate or chromic acid solution to stabilise the surface, followed by application of a very thin film of oil to assist lubricity in subsequent can-making processes.

Finally, after automatic and visual inspection, the strip is re-coiled or cut into sheets for packing and despatch.

There are two differing engineering designs used for electrolytic tinplate lines involving vertical or horizontal plating sections respectively. A vertical plating section (Fig. 87) is associated with Ferrostan and other acid processes and with alkaline tinplate lines. In this system the strip travels through the electrolyte in a series of vertical plating tanks, passing between tin anodes which hang on each side of it and both strip surfaces are plated simultaneously. A horizontal plating section (Fig. 88) is usually associated with the Halogen process (see above). In this process the tin anodes rest on the bottom of the plating tanks and the strip passes through the electrolyte above the anodes. One surface of the strip is plated in the first series of tanks and the other in the reverse pass through the second, upper series of tanks. Both types of line are today capable of operating at speeds up to 600 m/min.

Regardless of the production method, all tinplate is produced to the same technical specifications.

As stated above, relatively small amounts of tinplate are still produced by the older, hot-dipping method. In this the cold reduced steel, cut into sheets, is

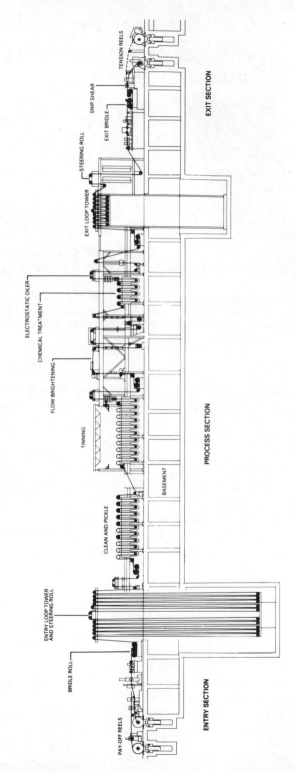

TENSION REELS

SNIP SHEAR

EXIT BRIDLE

STEERING ROLL

EXIT LOOP TOWER

ELECTROSTATIC OILER

CHEMICAL TREATMENT

FLOW BRIGHTENING

TINNING

CLEAN AND PICKLE

ENTRY LOOP TOWER
AND STEERING ROLL

BRIDLE ROLL

PAY-OFF REELS

EXIT SECTION

PROCESS SECTION

BASEMENT

ENTRY SECTION

Fig. 87 – Diagram of complete typical acid electrolytic tinplate line (Ferrostan process) with vertical plating cells.

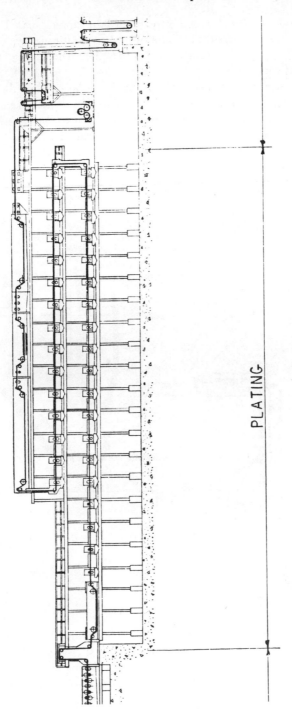

PLATING

Fig. 88 – Diagram of plating section of horizontal electrolytic tinplate line (Halogen process).

given a light acid pickle and passed into the hot-dip tinning unit (Fig. 89). This consists of a temperature-controlled bath of molten tin through which the sheets are conveyed via a series of submersed, driven rollers. The working temperature varies somewhat with the process, but is in the range 250 to 350 °C. Sheets enter the bath through a layer of molten flux, usually a mixture of fused zinc and other chlorides, and leave the 'tinpot' vertically upwards via a layer of molten palm oil which floats on the exit side of the bath and provides a lubricant film on the sheet. After tinning the sheets are inspected and packed.

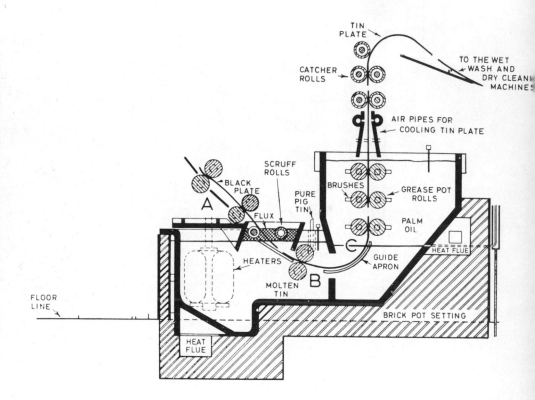

Fig. 89 – Diagram showing principle of unit for producing hot-dipped tinplate sheets.

Coating thickness in hot-dipped tinplate is controlled by the setting of the gap and the pressure between the exit pairs of rollers, assisted by the presence of the palm oil. It is not possible to produce such thin coatings — and hence such an economical product — as can be achieved by electrolytic tinning. This is the principal reason for the decline in hot-dipped tinplate production. A greater thickness of the compound layer (FeSn$_2$) is formed at the steel surface (Fig. 90) than in flow-melted electrolytic tinplate because of the longer time in contact with molten tin and the higher reaction temperature.

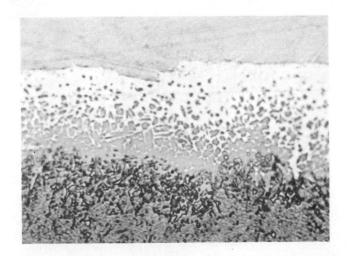

Fig. 90 – Taper microsection of hot-dipped tinplate used to improve resolution of the intermetallic compound layer (FeSn$_2$). (Etched in Picral.) (Magnification: horizontal × 2000; vertical × 20 000.)

## 9.3 PROPERTIES OF TINPLATE

The outstanding properties of tinplate are strength, formability, solderability, corrosion resistance — especially its resistance to attack by food products — non-toxicity, lacquerability and good appearance. It is this combination of properties which has led to its widespread use as a packaging material as well as in light engineering applications.

The mechanical properties of tinplate are essentially those of the steel base because the coating is so adherent that it will conform to any mechanical deformation that the steel will withstand without flaking or cracking.

The formability of tinplate is normally classified by a system of temper classifications. These are based on the Rockwell Superficial Hardness values, but the classifications embrace a range of properties. The usual range of temper classifications is set out in Table 23.

The range of coating masses has been defined previously in Table 21 and the common range of sheet thicknesses is in steps of 0.01 mm from 0.15 to 0.49 mm.

The manufacturer achieves the required mechanical properties by control of the various rolling and annealing procedures described above. Since these are independent of the coating process, it follows that any steel grade may be provided with any coating grade. This results in an enormous range of material properties covered by the generic name of tinplate.

### Table 23

Tinplate temper classifications

| Designation | | Forming characteristics | Approx. UTS $(kg/mm^2)$ |
|---|---|---|---|
| ISO | ASTM | | |
| T50 | T1 | Very ductile; suitable for severe draws, multistage draws (Cones, nozzles, intricate and deep-drawn containers and components) | 33 |
| T52 | T2 | Moderate drawing quality but having some measure of stiffness (Lid rings, closures, screw caps, drawn containers, e.g. fish cans) | 35 |
| T57 | T3 | General purpose quality (Can ends and bodies, large diameter closures, crown corks) | 37 |
| T61 | T4 | General purpose but additional stiffness (Stiffer bodies and ends and larger containers) | 41.5 |
| T65 | T5 | Stiff quality (End stock, large diameter containers, vacuum cans) | 45 |
| T70 | T6 | Very stiff quality (Beer cans and pressure-can ends) | 53 |
| CA61 | T4CA | Stamping quality, continuously annealed (All purposes requiring moderate drawing quality) | 41 |
| CA65 | T5CA | General purpose container stock, continuously annealed | 42.5 |
| CA70 | T6CA | Very stiff quality, continuously annealed (Beer can and pressure-can ends) | 53 |
| *DR550 | DR8 | Very stiff, directional (Round can ends and bodies) | 56 |
| *DR620 | DR9 | Very stiff, directional (Round can ends and bodies) | 63 |
| *DR660 | DR9M | Very stiff, directional (Beer cans and pressure-cans) | 67 |

* Proposed ISO designation for double-reduced grades.

## 9.4 APPLICATIONS OF TINPLATE

The principal use of tinplate is in the manufacture of rigid containers, chiefly for the food and beverage industries but also for a variety of non-food packs such as paints, oils and chemicals (Figs. 91 and 92). It is also used by the packaging industry in the form of caps and closures for glass containers and composite packages. In total some 90 per cent of tinplate goes into the container industry, with the rest being used in various light engineering applications. Much of the tinplate used is decorated or lacquered as flat sheets before being fabricated into articles for use.

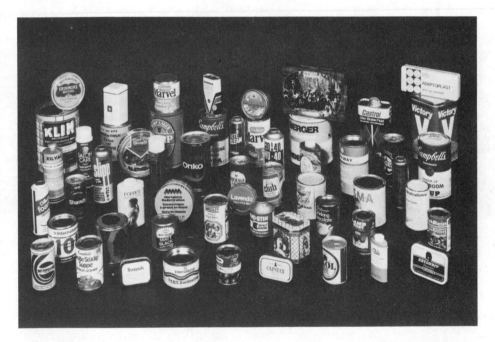

Fig. 91 – A wide variety of consumer goods are packaged in tinplate cans varying greatly in shape and size.

The traditional tinplate can is made from three pieces of tinplate, a body and two ends. The cylindrical can body is formed from a rectangular blank which has hooks formed on two opposing edges (see Fig. 53). After the body has been formed round a mandrel, the hooks are engaged and flattened and then soldered to form an impervious seam. Subsequently the cylindrical body is flanged at both ends and one end is mechanically seamed on using a plastic sealing compound to ensure a perfect seal. The other end is similarly seamed on by the packer after the can is filled and prior to sterilisation. A modern can-making line can operate at over 800 cans/min.

Fig. 92 – A selection of large drums and pails made from tinplate sheet.

Relatively new alternative methods of can-making aim at producing a two-piece can in which the body and one end are integral, thus eliminating two of the seams. The most important method is that of drawing and wall ironing (D & I). In this manufacturing process the starting material is a circular tinplate blank stamped from strip or sheet. The blank is initially formed into a shallow cup and then fed into an ironing press in which the cup wall is thinned and lengthened by forcing the cup, carried on a punch, through a series of successively smaller diameter ironing rings or dies (see Fig. 93). The resultant can body wall is some 30 to 50 per cent thinner than the starting material and considerably stiffer and stronger due to the work-hardening of the steel. Although soluble oils are used in the wall-ironing process, the tin coating is extremely important in assisting the lubrication, to the extent that it is not economically advantageous to use uncoated steel in this process.

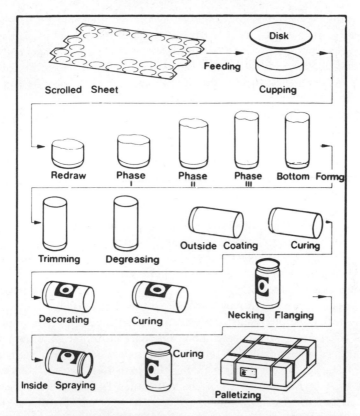

Fig. 93 – Manufacture of drawn and wall-ironed tinplate can – Top, flowsheet –
bottom, stages in the process.

This can-making procedure imposes severe requirements on the tinplate base quality. It must be clean, to avoid an unacceptable level of wall splits or flange cracks; composition must be controlled to give correct temper and formability; drawability and earing can influence operational performance, e.g. stripping from punches; the number of pinholes and welds must be controlled; gauge control within strict tolerances is essential.

A second method of producing a two-piece can is by single or multi-stage drawing. Shallow drawn containers have been used for many years, but recent multi-stage drawing processes, known as draw-redraw (DRD), are extending the range of sizes and applications for these types of can.

In this process, the can walls are not substantially thinned and the containers have sufficient strength to withstand processing, so that they are suitable for containing foodstuffs such as tuna fish, fruit, vegetables and meats. The DRD process has lower installation costs than drawing and ironing. In the draw-redraw process, can bodies are drawn from lacquered flat tinplate in several stages, the number of which depends on the height of the container required. Sizes are best restricted to those with small height to diameter ratios.

Welding as an alternative to soldering for securing side seams has been making rapid strides and is now a widely accepted can-making technique (Fig. 94). The technique of welding side seams on can bodies involves producing a series of overlapping weld nuggets along the length of the seam, using a suitable alternating current supply and copper electrodes. A problem with welding tinplate using copper electrodes is that as the tin melts it tends to alloy with the copper, thus altering the electrical characteristics. This has been overcome by using a copper wire as an intermediate electrode material. The wire is threaded over the rims of two electrode wheels and makes direct contact with the tinplate surface; as the tin melts, it is carried away on the wire and fresh copper is available for further welding.

On the bodymaker line, a tinplate blank is formed into a cylinder with suitable overlaps, and firmly held in position during passage through the unit. The welding electrodes (wheels carrying copper wire as described previously) are situated one above the other. The cylinder moves in such a way that the lower electrode is inside it; the seam thus lies between the two electrodes. When the current is passed between the two electrodes a series of spot welds is produced as the container moves, these constituting a continuous weld. Various commercial processes have been developed offering high welding speeds and a progressive reduction in the amount of overlap at the seam. Strong, hermetically sealed side seams can be made at high speeds. There is no risk of solder contamination of can contents and very thinly tin coated steel can be used. Lacquering procedures for welded three-piece cans are similar to those for the soldered cans. Beverage cans and aerosol containers have been manufactured by this method for some time and it is being increasingly used for food cans. Larger containers, such as drums, are frequently welded.

Fig. 94 — Side-seam welding of tinplate cans using the 'lost' copper wire technique which protects the wheel electrodes from erosion by the tin coating.

Ends for two- and three-piece cans are usually stamped from scrolled sheets, i.e. sheets sheared in a pattern which minimises scrap loss by allowing staggered rows of circular ends to be stamped. High speed presses for making can ends stamp the end into the required profile. Leaving the stamp press, the ends are fed into a curling unit and then to a machine which sprays a sealing compound around the inside of the curled edges; this ensures an hermetic seal when the closure is seamed on the can body. The ends are cured for 2–3 minutes. Bottom ends meet three-piece can bodies in can-making plants; top ends meet both two-piece and three-piece cans at the end of the filling line.

In addition to ends for cans, tinplate is used to make bottle closures, e.g. "crown corks" with their characteristic corrugated rims and cork or plastic pad as sealant, ensuring a tight hermetic seal when applied to a bottle.

A wide range of other glassware closures are produced in pre-decorated tinplate. Pre-formed screw caps are produced on special presses from decorated and lacquered tinplate and contain a plastic liner to ensure an hermetic seal. "Twist-off" closures are also made from tinplate by stamping from sheet.

The non-toxic, clean nature of tinplate makes it ideal for food contact applications and it is used to fabricate graters, cake tins, icers and bread bins. The strong fireproof nature of tinplate and its ease of decoration have lead to its use for waste-paper bins. Bodies of vacuum cleaners, reflectors of electric fires, cigarette lighters, children's toys, ferrules of paint brushes, and brightly coloured badges with soldered pins, are examples of the wide range of uses. An interesting application in Germany has been protective containers for bottles carrying chemicals. Tests have established that even when the container is dropped, the bottle inside will not break.

*Engineering applications*
The excellent fabricability of tinplate, its solderability, attractive appearance, and corrosion resistance, have led to a diversity of other uses besides packaging. Although none of these individual uses represents a large tonnage outlet, together they account for about 10% of total tinplate consumption. The starting point may be continuous coil, sheets, narrow strip or blanks of various shapes and sizes; the thickness, coating mass, temper and finish will be selected to match the intended end use. The wide range of grades in which tinplate is available, is an asset in this respect.

Around one million tonne of tinplate are used each year worldwide for non-packaging applications. In the automobile industry, tinplate is used for gaskets, in oil and air filters, for hydraulic systems and for oil-immersed coils as well as for a range of accessories such as ashtrays and mirrors, often overplated with another metal.

Tinplate is also employed widely in the electrical and electronics industries as cases for radios and amplifiers, for chassis and as shielding 'cans' for capacitors, relays and other components where corrosion resistance and solderability are required. The metal shielding for 'leakproof' batteries is made from tinplate. A number of domestic electrical appliances, such as reflector fires, use tinplate as a material of construction.

Many familiar objects such as portable butane cylinders and stoves, display signs and office accessories are also representative of the wide range of engineering uses for a versatile material.

Fig. 95 – An example of some non-package uses of tinplate. Courtesy: A. Tatham
& Sons Ltd.

# FURTHER READING

1.  Hoare, W. E., Hedges, E. S. and Barry, B. T. K., *The Technology of Tinplate,* Edward Arnold, London, 1965.
2.  *Tin versus Corrosion,* International Tin Research Institute Publication No 510.
3.  *Making, Shaping and Treating of Steel,* United States Steel Corporation, 7th edn., 1957.
4.  Kuntze, H. A., Siewert, J. H. and Bielefeld, W. F. H., Electrolytic strip tinning from 1934 to 1974 in the Andernach tinplate works of Rasselstein AG, *Stahl und Eisen,* Vol. 95, No. 4, p. 129, 1975.
5.  Johnson, W. R., The development of U.S. Steel Corporation's electrotinning process, *Stahl und Eisen,* Vol. 95, No. 7, p. 301, 1975.

6.  Smith, E. J., Swalheim, D. A. and McGraw, L. D., Development of the Halogen Process for electrolytic tinning of wide strip at high speeds, *Stahl und Eisen,* Vol. 95, No. 8, p. 355, 1975.

7.  *Guide to Tinplate,* International Tin Research Institute Publication No. 622.

# 10

# Tin Chemicals

This monograph is principally concerned with the properties and applications of tin from the metallurgical standpoint. Nevertheless, tin is capable of forming a wide range of chemical compounds and many of these are of industrial importance. For this reason, no book on tin would be complete without a chapter devoted to its chemical characteristics.

In its principal chemical reactions, tin can exhibit two valencies or oxidation states. Thus it is capable of forming divalent or stannous ($Sn^{2+}$) and quadrivalent or stannic ($Sn^{4+}$) compounds. The oxides of tin are also amphoteric, which means that tin can act as both acid and base in the formation of its compounds. Thus there are two further series of compounds possible, the stannites (divalent) and stannates (quadrivalent). Tin can also form organometallic compounds, often called 'organotins', in which from one to four carbon atoms may be bonded directly to a tin atom.

A more modern and systematic nomenclature for tin compounds refers to the oxidation state of the metal in a particular compound. Thus stannous and stannic compounds are classified as tin(II) and tin(IV) compounds respectively.

The industrial applications of tin chemicals are widespread and in many cases highly specialised. A major use is for metallic plated protective and decorative coatings (Chapters 8 and 9), but tin compounds also find uses in a vast number of totally unrelated fields, from wood preservation to ceramics, from toothpaste to plastics. This chapter summarises the more commercially important compounds and their applications.

## 10.1 INORGANIC COMPOUNDS OF TIN

### Tin(II) – Stannous compounds

Tin in the II+ oxidation state forms a large number of compounds. However, these stannous or tin(II) compounds are rather unstable in dilute aqueous solutions; they are susceptible to oxidation (to the IV+ state) and furthermore over a period of time hydrolysis can occur, with the formation of a hydrated tin(II) oxide or oxyhalide. These difficulties explain why tin(II) compounds have been comparatively little studied until recent years.

The most important industrial applications of stannous salts, in terms of volume, are in aqueous electrolytes for the plating of tin and tin alloys.

A series of divalent stannite compounds can be formed in which the tin atom is part of a complex anion. The stannites have no commercial value; indeed, the formation of stannites during electrodeposition of tin from alkaline electrolytes is a hindrance to efficient operation of the process (Chapter 8).

### Tin(II) Oxide, SnO

Stannous oxide is a stable, blue-black, crystalline product with a high metallic lustre. It is generally made by precipitating a stannous oxide hydrate ($SnO.nH_2O$) from a solution of stannous chloride by treating it with alkali. Heating the solution to near the boiling point, at carefully controlled pH, converts the white hydrate to the oxide. Tin(II) oxide burns in air, with incandescence, to form stannic oxide.

Stannous oxide readily reacts with mineral acids and with organic acids and is primarily used as an 'intermediate' in the manufacture of other tin compounds. It also finds use in the glass industry in the preparation of gold—tin and copper—tin ruby glass.

Stannous oxide dissolves in sodium hydroxide or potassium hydroxide solutions to form stannites, probably containing the $SnO_2^{2-}$ ion. These stannite solutions decompose readily to alkali metal stannates and tin.

### Tin(II) Hydroxide, Sn(OH)$_2$

Tin(II) hydroxide, stannous hydroxide, is obtained as a white, gelatinous precipitate by adding an excess of an alkali or an alkali carbonate to an aqueous solution of a stannous salt. Due to its nature the precipitate is difficult to filter. When freshly prepared, tin(II) hydroxide is white, but on standing it slowly absorbs oxygen from the air, being converted to hydrated stannic oxide.

Stannous hydroxide is only very slightly soluble in water, but is readily soluble in acids to form tin(II) salts. It reacts with aqueous solutions of sodium or potassium hydroxides to form soluble stannites; however this reaction does not occur with aqueous ammonia, nor with alkali carbonates.

### Tin(II) Chloride, SnCl$_2$

Tin(II) chloride, stannous chloride, is a white solid with a specific gravity of 3.95 at 25 °C. It crystallises in the monoclinic form as the dihydrate $SnCl_2.2H_2O$. When heated this salt gives off water, melts at 246.7 °C and boils at about 605 °C.

Anhydrous stannous chloride can be prepared directly by heating tin in hydrogen chloride gas or by direct chlorination. The hydrated form may be prepared by dissolving either tin metal or stannous oxide in warm hydrochloric acid. Tin($\text{II}$) chloride crystallises from solution as the dihydrate $SnCl_2.2H_2O$. However, dilute aqueous solutions tend to hydrolyse and to oxidise in air, to precipitate a basic tin chloride

$$SnCl_2 + H_2O \rightleftharpoons Sn(OH)Cl + HCl \;.$$

As will be seen from the equation, this hydrolysis, which is reversible, can be prevented by addition of excess HCl.

Solutions of tin($\text{II}$) chloride are strongly reducing and in fact find extensive use as reducing agents in laboratory procedures. Among the reactions for which stannous chloride is employed are for the reduction of ferric salts to ferrous, of cupric to cuprous, and the reduction of mercuric salts first to mercurous salts and finally to mercury. Tin($\text{II}$) chloride is also a useful reducing agent in certain organic chemical reactions. In its use as a reducing agent, stannous chloride is itself oxidised to the stannic form $SnCl_4$.

Neutralisation of tin chloride solutions with alkali causes stannous oxide or a metastable hydrate to be precipitated; addition of excess alkali results in the formation of a stannite.

There are numerous applications of tin($\text{II}$) chloride, but some of the most important are in plating applications (Chapters 8 and 9). The Halogen, or horizontal acid process for the manufacture of electrolytic tinplate employs a patented electrolyte containing an aqueous solution of stannous chloride and alkali metal fluorides. Stannous chloride is also a component of the plating bath for electrodepositing tin—nickel alloy coatings.

Tin($\text{II}$) chloride finds use as a sensitiser, both in silvering mirrors and in the plating of plastics. In producing mirrors or other reflective surfaces, the surface to be treated is sensitised with a solution of stannous chloride for up to one minute. This is washed off with running water and leaves a surface which will subsequently take a uniform, adherent film of silver. The plating of plastics often involves a sensitisation stage with stannous chloride, followed by activation with a noble metal solution, such as palladium chloride, which serves to catalyse subsequent deposition of electroless copper.

Considerable quantities of stannous chloride are used to stabilise perfume in toilet soaps. Other applications include tin coating of sensitised paper, as an anti-sludge agent for oils, and as an additive to drilling muds in the oil industry.

Tin($\text{II}$) Fluoride, $SnF_2$

Tin($\text{II}$) fluoride, stannous fluoride, exists as opaque, white, lustrous crystals, melting at 219.5 °C. It is readily soluble in water and, like stannous chloride, dilute solutions hydrolyse unless stabilised with excess acid. It is made commercially from stannous oxide and aqueous hydrofluoric acid, or by dissolving metallic tin in anhydrous or aqueous hydrofluoric acid.

The principal application of stannous fluoride is in toothpaste and in dental preparations, for preventing demineralisation of teeth. In 1960 the American Dental Association officially recognised the value of a commercial fluoride-containing toothpaste in preventing tooth decay and since then many commercial toothpastes containing stannous fluoride have been marketed. Solutions of stannous fluoride for topical application to teeth are also used by many dentists to prevent development of dental caries.

Other tin(II) halides exist, but have no significant commercial application.

### Tin(II) Bromide, $SnBr_2$

Tin(II) bromide, stannous bromide, is a light yellow salt, crystallising in the rhombic system as hexagonal columnar crystals. It has a melting point of 215.5 °C and, like the chloride and fluoride, is readily soluble in water.

Stannous bromide has been prepared by heating tin metal in an atmosphere of hydrogen bromide. An oily liquid condenses close to the hot zone and, on cooling, this freezes to the form which is solid at room temperature.

Tin(II) bromide readily hydrolyses and it forms addition compounds with ammonia and the bromides of other metals. It behaves similarly to the chloride but has no major industrial applications.

### Tin(II) Iodide, $SnI_2$

Tin(II) iodide, stannous iodide, is a red, crystalline substance with a melting point of 316 °C. It is sparingly soluble in water, but is soluble in hydrochloric acid and in potassium hydroxide. It is also soluble in hydriodic acid or iodides to form $HSnI_3$ or its salts.

When finely divided tin is heated with iodine, a mixture of Sn(II) and Sn(IV) iodides is produced. They can be separated by sublimation, since the stannic salt volatilises at 180 °C leaving the Sn(II) iodide. Stannous iodide may also be prepared by reducing stannic iodide with metallic tin in a sealed tube under prolonged heating at 360 °C.

### Tin(II) Sulphate, $SnSO_4$

Tin(II) sulphate, stannous sulphate, is a white, crystalline powder, thermally stable to about 360 °C and soluble in water. One manufacturing method is to react granulated tin with excess sulphuric acid (specific gravity 1.53) for a period of days at 100 °C until all reaction has ceased. Water is added and the aqueous solution evaporated *in vacuo*. Excess acid can be removed with alcohol. The sulphate can also be made from stannous oxide and sulphuric acid.

A major use for stannous sulphate is in tin plating processes for both static bath plating and for high-speed continuous plating lines (Chapters 8 and 9). A tin(II) sulphate-based electrolyte is stabilised against atmospheric oxidation by the addition of phenolsulphonic or cresolsulphonic acids.

One of the most widely used manufacturing processes for the production of electrolytic tinplate makes use of a bath based on stannous sulphate. This is the process developed by the US Steel Corporation for the production of Ferrostan tinplate. The patented bath differs slightly from the composition used in static plating in order to permit high rates of deposition on steel strip which may be moving through the electrolyte at up to 600 m/min. Anodes of refined tin are used and, because of balanced anode and cathode efficiencies, the bath needs very little addition of stannous sulphate during operation, the required tin content of the bath being maintained by dissolution of the replaceable anodes.

### Tin(II) Sulphide, SnS

Tin(II) sulphide, stannous sulphide, is formed as a grey crystalline mass when tin is heated with sulphur, or as a brown precipitate when hydrogen sulphide is passed into an aqueous solution of a stannous salt. This latter reaction is made use of in analytical procedures.

Tin(II) sulphide is sparingly soluble in water but will dissolve in hot concentrated hydrochloric acid. Stannous sulphide will not dissolve in alkali sulphides (for example, ammonium sulphide) if these are perfectly free from excess sulphur, but dissolves readily in yellow ammonium sulphide (which contains excess sulphur) to form ammonium thiostannate

$$SnS + S + (NH_4)_2 S \rightarrow (NH_4)_2 SnS_3$$

### Tin(II) Fluoroborate, $Sn(BF_4)_2$

Tin(II) fluoroborate, stannous fluoroborate, is known only in solution, the solid not having been isolated. Solutions are prepared by dissolving stannous oxide in fluoroboric acid. The commercial solution contains 47 per cent stannous fluoroborate and is widely used in tin and tin–lead alloy plating. Some tinplate is produced using an electrolyte based on this compound (Chapter 9).

### Tin(II) Arsenate, $SnHAsO_4.\frac{1}{2}H_2O$

Tin arsenate, prepared by the reaction of stannous chloride with arsenic acid, is of commercial interest in view of its antihelminthic action, i.e. its activity against parasitic worms, in animals and birds. Many metallic arsenates are antihelminthic, but tin arsenate has been found to be the most effective.

### Tin(II) Pyrophosphate, $Sn_2P_2O_7$

Stannous pyrophosphate is an amorphous white powder, insoluble in water but soluble in sodium pyrophosphate and hydrochloric acid solutions. It is prepared from stannous chloride and sodium pyrophosphate and is used in several alkaline plating baths for the deposition of tin alloys. Alloys containing copper with varying amounts of tin can be deposited from a copper cyanide/stannous pyrophosphate bath by appropriate variation of bath composition and operating conditions.

### Other Tin(II) compounds

A number of tin(II) salts of organic acids have some industrial importance.

Tin($II$) Acetate, $Sn(C_2H_3O_2)_2$

Stannous acetate is a colourless crystalline salt, melting point 182 °C, which gives off a strong odour of acetic acid when exposed to humid air. A solution of the acetate may be obtained by metathetical reaction of lead acetate and stannous chloride or by dissolving stannous oxide in glacial acetic acid. It can also be made by dissolving metallic tin in glacial acetic acid provided oxygen is present to depolarise the surface. The dry salt can be obtained by precipitation from the aqueous solution with alcohol.

The acetate can be used as a catalyst in the high pressure hydrogenation of coal and has been employed industrially to promote dye uptake by fabrics.

Tin($II$) 'octoate' (Stannous 2-ethylhexoate), $Sn(C_8H_{15}O_2)_2$

Stannous 2-ethylhexoate is an industrially important tin chemical and is usually referred to as stannous 'octoate'. It may be made by reaction of stannous oxide with 2-ethylhexoic acid and is a clear, light yellow, somewhat viscous liquid. It is soluble in many organic solvents including silicone oils.

The major use of this chemical is as a catalyst and it performs broadly the same function as the organotin compounds dibutyltin dilaurate and dibutyltin diacetate in certain types of silicone elastomers and polyurethane foams (Fig. 96). These catalytic applications are described later under 'Organotins'.

Other stannous soaps derived from lauric, oleic and stearic acids are available in proprietary formulations for similar catalytic purposes.

Tin($II$) formate, $Sn(CHO_2)_2$

Stannous formate forms white, anhydrous crystals which decompose above 100 °C in air. It is obtained by treating stannous oxide with formic acid and has been patented as a catalyst for the hydrogenation of liquid fuels.

Tin($II$) oxalate, $Sn(C_2O_4)$

Stannous oxalate is a white, crystalline powder which decomposes at 280 °C. It is insoluble in water but soluble in a mixture of oxalic acid and ammonium oxalate solutions. It is an ideal catalyst for esterification reactions since it limits destructive side reactions which are responsible for degradation of esters at preparation temperatures. It is best employed in direct esterification reactions involving secondary alcohols or in trans-esterification or poly-esterification reactions.

Tin($II$) tartrate, $Sn(C_4H_4O_6)$

Stannous tartrate is a white crystalline powder, insoluble in water but soluble in dilute hydrochloric acid. It can be prepared from stannous chloride and potassium tartrate and is used in dyeing and in the printing of textiles.

**Tin($IV$) – Stannic compounds**

Tin in the $IV+$ oxidation state forms a large number of compounds, many of

Fig. 96 — Production of polyurthane foam using stannous octoate as the catalyst.

which are of major industrial importance. In addition to being manufactured on a fairly large scale, tin(IV) compounds have been the subject of much study and their properties are widely documented.

Some compounds, for example stannic oxide, have a long history of use in traditional industries such as the production of glass and ceramics. Other compounds, like the chloride, have recently found new uses in modern technical processes such as in strengthening treatments for lightweight glassware. The stannates, in which the tin atom is in a complex anion, are another group of tin(IV) compounds which find many uses both in electroplating and in the electrical industries.

### Tin(IV) Oxide, $SnO_2$

Tin(IV) oxide, stannic oxide (or tin dioxide), occurs naturally, crystallised in the tetragonal system, as the mineral cassiterite, which is the principal tin mineral (Chapter 2). The oxide can also exist in rhombic and hexagonal forms and is trimorphous.

Stannic oxide can be prepared industrially by a number of routes: (a) by blowing heated air over molten tin; (b) by atomising tin using high pressure steam and burning the finely divided metal; (c) by calcination of hydrated oxide. It can also be prepared by treating stannic chloride with steam at high temperature, by treating granulated tin with nitric acid at room temperature, or by neutralising stannic chloride with a base.

Tin(IV) oxide is a refractory material, i.e. it has an extremely high melting point, about 1600 °C. It is therefore frequently used in the ceramics industry as a pigment and as an opacifier in glazes (Fig. 97). Because of its insolubility in various glazes, stannic oxide serves as a base for colours such as chrome—tin pink and vanadium—tin yellow (with chromium and vanadium oxide respectively). When used as a glaze opacifier in additions of 4 to 5 per cent, it gives a pure white glaze.

Stannic oxide has a hardness of 6 to 7 (on the Mohs' scale) and finely divided and graded stannic oxide powder is used as a grinding medium for polishing marble and granite. In this form it is sometimes known as 'putty powder'. Substantial quantities are used industrially and smaller amounts by hobbyists for polishing pebbles and semi-precious stones to make costume jewellery. It has an important industrial application as a polishing medium in the production of sewing needles.

A thin and completely transparent film of tin(IV) oxide, deposited on glass surfaces to a thickness of about 0.1 $\mu$m, has been found to increase substantially the strength of the glass and to increase its abrasion resistance. This coating process is usually based on treatment with stannic chloride or dimethyltin dichloride. Thicker coatings of stannic oxide on glass confer electrically conducting properties to the coated surface. Although usually classed as a semiconductor in structure, a stannic oxide coating possesses specific electrical resistivity and

Fig. 97 – Tin oxide based pigments in glazed tiles used in a London station.

conductance properties akin to those of metals. Industrial applications utilising the electrical properties of thin stannic oxide films on glass or ceramics include precision electrical resistors having high thermal stability and reliability, special lighting fixtures (such as illuminated signs), and de-icing glass (for example, for the windows of aircraft).

### Tin(IV) oxide and hydrates

Hydrated stannic oxides of variable water content can be obtained by the hydrolysis of stannates.

An hydrated tin oxide may be precipitated as a flocculent white mass by acidifying a solution of sodium stannate. If this precipitate is washed free of water-soluble ions and peptised with potassium hydroxide, a colloidal solution is produced which is stable below 50 °C for long periods.

Hydrated stannic oxide has been shown to hold good promise in the field of ion exchange and also has potential as a catalyst in certain oxidation reactions, either when combined with another metal oxide (for example, copper oxide) or as the active receptor for a precious metal catalyst (such as palladium).

### Tin(IV) Chloride, $SnCl_4$

Tin(IV) chloride, stannic chloride or tin tetrachloride, is a colourless liquid which fumes in moist air and becomes turbid as a result of hydrolysis which produces finely divided hydrated tin oxide or basic chloride.

Tin(IV) chloride freezes at −30.2 °C, boils at 114 °C and has a specific gravity of 2.23 at 20 °C. The simplest and most usual method of preparation is by direct chlorination of metallic tin

$$Sn + 2Cl_2 \rightarrow SnCl_4$$

During chlorination some tin(II) chloride may also be formed, but the tetrachloride is easily separated by evaporation. In the absence of water, tin(IV) chloride is inert to steel and may be shipped in plain steel drums of special design. However, prolonged contact with skin can cause burns and suitable precautions should be taken when handling stannic chloride. Although dry stannic chloride is an almost perfect electrical insulator, traces of water make it weakly conducting.

Tin(IV) chloride reacts with water, with evolution of heat, to form a pentahydrate, $SnCl_4.5H_2O$, which is a white, crystalline solid. The pentahydrate is deliquescent and is very soluble in water or in alcohol. It is stable from 19 to 56 °C and is often used in place of the anhydrous chloride when the presence of water is not objectionable, since it is easier to handle.

An important use for tin(IV) chloride is as a starting material for the manufacture of organotin compounds (*q.v.*). Other applications of stannic chloride include several relatively small uses such as in the manufacture of fuchsin, of colour lakes and ceramics, for weighting of silk and for stabilisation of perfumes in toilet soap. It has also been used for many years as a mordant in the dyeing of silks.

A relatively new use which is growing in industrial importance is for the surface treatment of glass containers to increase their strength. In this process freshly formed glassware is passed through an annealing oven in which is maintained an atmosphere containing stannic chloride vapour. This chloride coats the glass and at the annealing temperature decomposes, leaving a deposit of tin(IV) oxide on the surface of the glass. This film is completely transparent but so strengthens the glass and improves its abrasion resistance that glassware treated in this way can be made considerably thinner and hence lighter in weight, with consequent economic advantages. The process is widely used in the production of bottles and other glass containers and for domestic glassware.

Stannic chloride forms double salts with alkali chlorides, for example $(NH_4)_2SnCl_6$, known as ammonium chlorostannate or ammonium stannichloride.

Other stannic halides are known to exist but are not of major commercial importance.

## Tin(IV) Fluoride, $SnF_4$

Stannic fluoride may be formed by direct action of fluorine on tin at 100 °C. It can also be produced by the addition of tin(IV) chloride to anhydrous hydrofluoric acid. Tin(IV) fluoride is a white crystalline solid which sublimes at 750 °C. It is very hygroscopic.

## Tin(IV) Bromide, $SnBr_4$

Stannic bromide is a white, fuming, crystalline substance. It can be formed directly by burning tin in an atmosphere of bromine. The bromide has a melting point of 31 °C and is fairly stable when heated. It can be sublimed without decomposition.

## Tin(IV) Iodide, $SnI_4$

Stannic iodide is a stable, red, crystalline solid which can be prepared directly by heating tin with iodine. The reaction produces a mixture of tin(II) and tin(IV) iodides which can be separated by heating above 180 °C, at which temperature the stannic iodide volatilises and is sublimed. The stannous salt is unaffected at this temperature.

## Tin(IV) Sulphide, $SnS_2$

Tin(IV) sulphide, stannic sulphide, exists as golden-yellow scales, specific gravity 4.51. It is usually prepared technically by dry methods, for example by heating tin metal with elemental sulphur and ammonium chloride. It is precipitated by hydrogen sulphide from weakly acid solutions of tin(IV) salts and dissolves readily in alkaline sulphide solutions, forming soluble thiostannates such as

$(NH_4)_2SnS_3$. Because of its lustrous colourful appearance, the sulphide has been known for many years as a bronzing agent for treating wood, gypsum, etc. It is referred to in antiquity as 'mosaic gold' and is now commercially available as a bronzing agent. It also finds use as a pigment and has been studied as a semiconductor.

### Tin(IV) Sulphate, $Sn(SO_4)_2.2H_2O$

Tin(IV) sulphate, stannic sulphate, may be formed by the solution of stannic hydroxide in dilute sulphuric acid or by the action of oxidising agents on stannous salts. Stannic sulphate is hydrolysed by water.

### Tin(IV) Phosphate, $Sn_3(PO_4)_2$

Tin(IV) phosphate, stannic phosphate, is obtained by adding sodium phosphate to a solution of tin(IV) sulphate containing a little $H_2SO_4$. It is a white amorphous solid which is soluble in dilute mineral acids and in alkalis.

Stannic phosphate gel is prepared by reacting tin(IV) chloride with sodium phosphate and obtaining a resultant precipitate. It is an ion exchanger and has possible industrial potential in this application.

The hypophosphate, $SnHPO_4$, is obtained as a crystalline solid by dissolving tin in phosphoric acid.

### Tin(IV) Vanadate

Stannic vanadate is an oxidation catalyst and it has been patented for this action for a number of applications including the oxidation of sulphur dioxide and for the oxidation of hydrocarbons. It also finds limited use in certain inorganic pigments for ceramics.

### Stannates

As has been stated, tin and its compounds are amphoteric and therefore the tin ion can exhibit both acidic and basic properties. Stannic acid is generally regarded as an hydrated stannic oxide and its formula is usually considered to correspond to $H_2SnO_3$.

Whilst stannic acid itself is of little importance, it forms a large series of salts, the stannates, two of which, sodium stannate $Na_2Sn(OH)_6$ and potassium stannate $K_2Sn(OH)_6$, are important industrial chemicals.

Sodium or potassium stannate can be prepared by fusing stannic oxide with sodium hydroxide or potassium carbonate respectively, followed by leaching. The alkali stannates are often obtained industrially from secondary tin by recovery from detinning solutions.

The most important applications of these compounds are for plating tin and its alloys (Chapter 8). Electrolytes based on sodium stannate are commonly used in Europe whilst in the USA the potassium stannate bath is the favoured process.

The original electrolytic tinplate lines used stannate solutions but these have been largely superseded by acid electrolytes. Alkaline plating is still widely used for coating finished components.

Bismuth stannate is a light-coloured, crystalline powder of formula $Bi_2(SnO_3)_3.5H_2O$. The approximate temperature of dehydration is 140 °C. Bismuth stannate and barium titanate are often combined to produce a ceramic capacitor body of uniform dielectric constant ($K$ value) over a substantial temperature range.

A large number of different stannates are available commercially for dielectric applications. A few stannates are also used in making pigments and lakes; these include chromium(III) stannate and cobalt(II) stannate.

## 10.2 ORGANOTIN COMPOUNDS

Organotin compounds are defined as compounds in which at least one tin–carbon bond exists. The great majority of organotin compounds have tin in the IV+ oxidation state. Tin–carbon bonds are in general weaker and more polar than those formed in organic compounds of carbon, silicon or germanium, and organic groups attached to tin are more readily removed. This higher reactivity does not, however, imply an instability of organotin compounds under ordinary conditions.

Four series of compounds are known, categorised as $R_4Sn$, $R_3SnX$, $R_2SnX_2$ and $RSnX_3$. In the organotin chemicals of commercial importance, R is usually a butyl, octyl or phenyl group and X is commonly chloride, fluoride, oxide, hydroxide, carboxylate or thiolate (Fig. 98).

Organotin chemicals have been known for over one hundred years; the chemist Frankland prepared diethyltin di-iodide in 1849. Although much significant work on organotins was performed in the next few decades, interest waned, and it was not until the late 1940s that renewed interest arose from realisation of the commercial potential of these compounds. The first patent covering the use of dialkyltin compounds as stabilisers for polyvinyl chloride was taken out in 1936, and in the mid-1940s organotin compounds were used for that purpose in the USA. The discovery in the 1950s that triorganotin compounds had pronounced biocidal properties led to a whole range of new uses for organotins. The later finding that dioctyltin compounds possess negligible mammalian toxicity has resulted in organotin stabilisers based on these latter compounds being used in PVC for packaging foodstuffs and especially for making plastic bottles (Fig. 99).

The increase in commercial importance of the organotins is reflected in the tonnage consumption figures. In 1950 the annual production was around 50 tonnes. There was a sharp rise after 1955 and by 1960 annual use was around 2000 tonnes. By 1968 usage had grown to 10 000 tonnes and today is about 30 000 tonnes per annum.

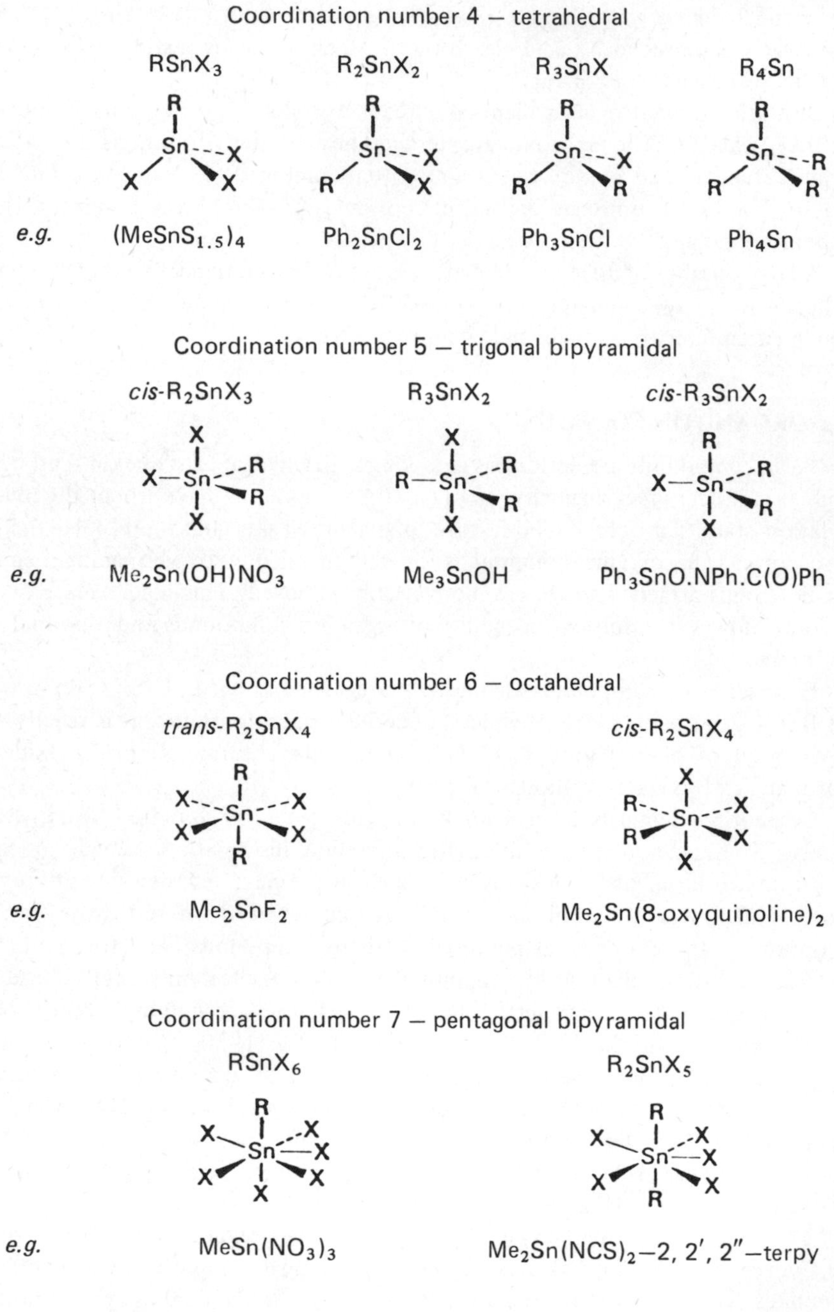

Coordination number 4 — tetrahedral

| $RSnX_3$ | $R_2SnX_2$ | $R_3SnX$ | $R_4Sn$ |

*e.g.*  $(MeSnS_{1.5})_4$     $Ph_2SnCl_2$     $Ph_3SnCl$     $Ph_4Sn$

Coordination number 5 — trigonal bipyramidal

*cis*-$R_2SnX_3$     $R_3SnX_2$     *cis*-$R_3SnX_2$

*e.g.*  $Me_2Sn(OH)NO_3$     $Me_3SnOH$     $Ph_3SnO.NPh.C(O)Ph$

Coordination number 6 — octahedral

*trans*-$R_2SnX_4$     *cis*-$R_2SnX_4$

*e.g.*  $Me_2SnF_2$     $Me_2Sn(8\text{-oxyquinoline})_2$

Coordination number 7 — pentagonal bipyramidal

$RSnX_6$     $R_2SnX_5$

*e.g.*  $MeSn(NO_3)_3$     $Me_2Sn(NCS)_2-2, 2', 2''-\text{terpy}$

Fig. 98 – Diagram showing structures of basic types of organotin compound.

Fig. 99 — Organotin-stabilised PVC bottles being filled with fruit squash. (Courtesy: Able Development Ltd.)

### Tetraorganotin compounds R₄Sn

The tetraorganotins correspond to the formula $R_4Sn$, where R is an organic group, usually an alkyl or an aryl group. Tetraalkyl and tetraaryl tin compounds are mainly important as intermediates in the preparation of other organotin compounds. Although some of these compounds are used commercially, industrial usage is not great at the present time.

The tetraalkyltins are colourless and the compounds of lower molecular weight are liquids at room temperature; higher molecular weight members of the series are low-melting solids.

The lower tetraalkyltins can be distilled at atmospheric pressure without decomposition and are soluble in the common organic solvents. Higher molecular weight substances can be dissolved only in solvents such as benzene, pyridine or chloroform. The tetraaryltins are solids with melting points above 179 °C.

Tetraorganotin compounds are regarded as possessing typical covalent bonds. They are stable in the presence of air and water and, although they are not highly sensitive to strong aqueous bases, cleavage of the tin—carbon bond occurs readily with halogens, hydrogen halides or strong aqueous acids.

Since they serve as starting points for the preparation of other organotins, the methods of preparation of the tetraorganotins are important, holding the key to efficient commercial manufacture of organotin compounds. Three of the four methods presently used for technical manufacture of organotin compounds involve the preparation of $R_4Sn$ compounds from $SnCl_4$ as a first step:

The Grignard process:

$$SnCl_4 + 4RMgCl \rightarrow R_4Sn + 4MgCl_2$$

The Wurtz method:

$$SnCl_4 + 4RCl + 8Na \rightarrow R_4Sn + 8NaCl$$

The aluminium-alkyl technique:

$$3SnCl_4 + 4R_3Al \rightarrow 3R_4Sn + 4AlCl_3 \ .$$

For the manufacture of tetraphenyltin, so far only the Grignard process appears feasible. The same process is being widely used on a large scale for making tetrabutyltin and tetraoctyltin. Although many improvements have been made to the Grignard manufacturing process, the necessity for using mixed solvent systems and the large volumes involved are technical limitations of the method.

The Wurtz method is practised on a manufacturing scale both in the USA and in East Germany for manufacturing tetrabutyltin. A simple hydrocarbon solvent can be used but again a large-volume system is involved. The aluminium-alkyl method is used in West Germany for manufacture of tetraalkyltins, particularly tetraoctyltin. No solvent is required but the complete transfer of all the alkyl groups bound to the aluminium requires the presence of certain complexing agents such as sodium chloride, ethers or tertiary amines.

Most organotin compounds of commercial interest are prepared from the tetraorganotins as a starting point. Treatment with stannic chloride is a widely used procedure for preparation of organotin halides, and redistribution reactions occur:

$$3R_4Sn + SnCl_4 \rightarrow 4R_3SnCl$$

$$R_4Sn + SnCl_4 \rightarrow 2R_2SnCl_2$$

$$R_4Sn + 3SnCl_4 \rightarrow 4RSnCl_3$$

Further redistribution can then occur by reacting with organotin halides:

$$R_4Sn + RSnCl_3 \rightarrow R_3SnCl + R_2SnCl_2 \ .$$

The relative ease and efficiency with which alkyl or aryl groups can be redistributed amongst the tin atoms is one of the most unusual features of organotin chemistry and is essential for economic production of organotin chemicals. From the halides other derivatives may then be obtained.

### Tetrabutyltin, $Sn(C_4H_9)_4$

Tetrabutyltin is a colourless, oily liquid with a distinct but not unpleasant odour. Its properties are shown in Table 24.

It can be prepared by reacting $SnCl_4$ with an excess of butyl magnesium chloride in toluene above 100 °C; metal chloride is removed with water and the tetrabutyltin can be distilled for further purification if required.

It has been claimed that optimum yield is obtained when a slurry of finely dispersed sodium in a hydrocarbon solvent reacts with butyl chloride and tin tetrachloride under carefully controlled temperature conditions.

The main commercial use is as a corrosion-inhibiting additive to lubricating oils of mineral hydrocarbon origin, which are normally corrosive towards bearing metal alloys. Its use has also been suggested in combination with $AlCl_3$ and certain transition metal chlorides as a catalyst for low pressure polymerisation of olefines. The organotin acts as an *in situ* alkylating agent for $AlCl_3$ and avoids the handling of extremely reactive organoaluminium compounds.

### Tetraphenyltin, $Sn(C_6H_5)_4$

Tetraphenyltin is a white, crystalline powder, slightly soluble in organic solvents at room temperature, more soluble at higher temperatures in aromatic solvents and insoluble in water. The principal physical and chemical properties are shown in Table 25.

**Table 24**

Physical and chemical properties of tetrabutyltin

| Compound | Physical form | Specific gravity at 20 °C | Molecular weight | Melting point (°C) | Boiling point (°C) | Solubility in Water | Solubility in Solvents |
|---|---|---|---|---|---|---|---|
| Tetrabutyltin | liquid | 1.05 | 347 | −97 | 145 | insol. | sol. most solvents |

**Table 25**

Physical and chemical properties of tetraphenyltin

| Compound | Physical form | Specific gravity at 25 °C | Molecular weight | Melting point (°C) | Boiling point (°C) | Solubility in Water | Solubility in Solvents |
|---|---|---|---|---|---|---|---|
| Tetraphenyltin | solid | 1.48–1.49 | 427 | 224–230 | >420 | insol. | slightly sol. at room temp., more sol. at higher temp. |

Fig. 100 — Lengths of joinery timber being withdrawn from a vacuum treatment chamber after impregnation with tributyltin oxide-based wood preservative.

It is best prepared by a Grignard reaction; if tetrahydrofuran is used as the solvent, the less expensive phenylmagnesium chloride can be used in place of the corresponding bromide. Chlorobenzene and magnesium are reacted in tetra-hydrofuran in the first stage and then $SnCl_4$ is added for the second phase of the reaction.

Aryl groups bonded to tin can be cleaved by mineral acids such as hydro-chloric acid; this makes them useful scavengers for traces of such acids. Tetra-phenyltin is used for this purpose in chlorinated dielectric fluids. Like tetra-butyltin, the tetraphenyltin compound can also be used in catalyst systems for olefine polymerisation.

## Triorganotin compounds $R_3SnX$

The triorganotins correspond to the formula $R_3SnX$, where R is an organic group, typically an alkyl or an aryl group. Triorganotin compounds form one of the most important classes of organotin chemicals in view of their wide commercial use as biocides. These applications result from the fact that organotin compounds containing three tin-to-carbon bonds exhibit high biocidal activity.

Whilst the nature of the organic group R strongly influences the biological properties of the compound, the inorganic group X appears to have no direct influence on the biocidal properties except through its effect on solubility and volatility. This means that a whole series of derivatives can be developed to meet the requirements of specific end uses.

Maximum biocidal activity is known to occur in compounds where the total number of carbon atoms in the three R groups lie in the range 9 to 12. Tributyltin compounds are almost ideal wood preservatives. They can be applied to timber by dipping, spraying or by vacuum impregnation methods (Fig. 100), a suitable organic solvent system being used. Provided normal industrial precautions are followed, the tributyltin compounds present no health hazards in application or use.

The growing commercial importance of this use is reflected in the fact that in the UK alone there are over 30 formulations containing organotins com-mercially available for timber preservation (Fig. 101). The National House-Builders' Registration Council requires that all wooden window joinery be given a preservative treatment, and a dip in organic solvent containing tributyltin oxide is considered suitable.

Tributyltin compounds are also widely used in anti-fouling paints for marine use. The hulls of boats which are in prolonged contact with water are susceptible to attachment by plant growth and adherence of marine organisms such as barnacles which can seriously interfere with the runnong efficiency of the vessel. Organotin compounds are finding increasing use in the special paints which protect the hulls against this type of fouling. The toxic agent is diffused through the paint film and slowly released to the water in contact with the hull, where it effectively kills off adjacent marine organisms.

Fig. 101 – Building timbers are commonly protected against wood-rotting fungi by impregnation with TBTO. (Courtesy: New Ideal Homes Ltd.)

The triorganotin compounds have special advantages for this application, since in addition to being highly active against a wide range of fouling species they are colourless, allowing light pastel shades to be used in paints, and give no corrosion problems even on aluminium hulls where paints containing certain other metals can cause galvanic corrosion.

For agricultural use one advantage of organotins is the fact that they eventually break down to non-toxic forms and so introduce no long-lasting residual toxicity. They have been used with success against many fungal diseases which cause damage to essential crops such as potatoes. Tests upon potato and sugar beet fields in many countries under different climatic conditions have established that the amount of residual organotin in the product never reaches the limits proscribed by public helath authorities.

Triorganotins are also effective miticides and a significant development was the discovery that some of these compounds have an antifeedant effect, making the plants treated with them unpalatable to the feeding insect at concentrations too low to be lethal.

Yet another application for these compounds is as bactericides. Tributyltin compounds are effective against many gram-positive bacteria, including the antibiotic-resistant staphylococci which are often the source of cross-infection in hospitals. Hospital tests in which the compounds were incorporated into waxes and polishes, sprays for walls and ceilings and in laundry washes, demonstrated a considerable reduction in staphylococci after systematic treatment. The compounds have also been used with success to prevent infective fungi and bacteria in socks, footwear and textiles. Commercial preparations usually contain both the organotin and another bactericide such as formaldehyde, or a quaternary ammonium salt.

Tributyltin compounds in quite small concentrations will prevent the formation of slime and fungal growths in water and have been used for this purpose in paper manufacturing mills and in closed circuit cooling water systems. Another application is in protecting coats of emulsion paint against mildew formation when the paint is subjected to moist, humid, environments. A very low concentration of the tributyltin compound is effective for this purpose.

One interesting development has been the incorporation of tributyltin compounds into rubber so that in contact with water, the organotin is slowly released to the water from the rubber surface. This leaching is accompanied by a redistribution of the organotin within the matrix so that more is made available at the surface, and this process continues until all the compound is eventually released to the water. Impregnated rubber has been used successfully as an anti-fouling coating in applications such as protection of sonar buoys which are moored in seawater for long periods. This principle also offers hope of being a potent weapon in the fight against bilharzia. Triorganotins can be used as molluscicides for killing the snails which carry the bilharzia, and one development has been the incorporation of organotins in pellets or strips which can be immersed in water so that the active ingredient is slowly released to the water at low concentrations which are toxic to the parasite but do not affect fish and plant life.

Two of the most important classes of triorganotin compound are the tributyltin and the triphenyltin derivatives. Both these types combine high biological activity with low mammalian toxicity. The principal physical and chemical properties of the more industrially important compounds are given in Tables 26 and 27 respectively. Details of the individual compounds are given below.

### Tributyltin oxide, $(C_4H_9)_3Sn.O.Sn(C_4H_9)_3$

Tributyltin oxide is a colourless liquid prepared by heating a mixture of tributyltin chloride and aqueous sodium hydroxide. In one patented manufacturing process, the chloride is refluxed with sodium chloride in ethanol.

Tributyltin oxide is widely used as a biocide in many of the applications mentioned above. It has the advantages of being compatible with a great many other biologically active compounds and its biocidal activity is maintained even in alkaline solutions and in dilute mineral acids at room temperature. Tributyltin oxide is sometimes used in adhesives for vinyl wallpaper and tiles to prevent bacterial decomposition of some of the adhesive constituents.

### Tributyltin chloride, $(C_4H_9)_3SnCl$

Tributyltin chloride is a colourless liquid prepared by heating tetrabutyltin with stannic chloride at 210 to 230 °C. It is used as a starting material for the production of a number of other tributyltins.

Tributyltin chloride has been used in USA as a rodent repellent in plastics-based covering for electrical cables. The coating is not lethal to rats (who do considerable damage to untreated cables) but has a repellent action and 95 per cent effectiveness has been claimed in tests.

### Tributyltin fluoride, $(C_4H_9)_3SnF$

Tributyltin fluoride is a solid in the form of white, fine, prismatic crystals melting at 240 °C with decomposition. It is insoluble in water but slightly soluble in organic solvents. Tributyltin fluoride finds use in anti-fouling paint formulations (Fig. 102).

### Tributyltin acetate, $(C_4H_9)_3SnOOC.CH_3$

Tributyltin acetate is a white, crystalline solid made by reacting tributyltin chloride with potassium acetate.

The major use for tributyltin acetate is in anti-fouling paint formulations. It has also been used as a stabiliser for halogen-containing polymers and as a catalyst in the production of polyurethane foams.

**Table 26**

Physical and chemical properties of some tributyltin compounds

| Compound | Physical form | Specific gravity at 20 °C | Molecular weight | Melting point (°C) | Boiling point (°C) | Solubility in Water | Solubility in Solvents |
|---|---|---|---|---|---|---|---|
| Tributyltin oxide | liquid | 1.17 | 596 | <−45 | 180(2 mm) | insol. | sol. |
| Tributyltin chloride | liquid | 1.2−1.3 | 325.5 | 30 | 210−214(10 mm) 142(15 mm) 172(25 mm) | insol. cold, hydrolyses hot | sol. most solvents |
| Tributyltin fluoride | solid | – | 309 | 240 | | insol. | slightly sol. most solvents |
| Tributyltin acetate | solid | 1.27 | 349 | 80−85 | | insol. | sol. benzene and methanol |
| Tributyltin benzoate | liquid | 1.19 | 411 | | 166−168(1 mm) | | |

**Table 27**

Physical and chemical properties of some triphenyltin compounds

| Compound | Physical form | Molecular weight | Melting point (°C) | Solubility in Water | Solubility in Solvents |
|---|---|---|---|---|---|
| Triphenyltin acetate | solid | 409 | 119−124 | insol. | slightly sol. in alcohols and aromatic solvents |
| Triphenyltin chloride | solid | 385.5 | 103−107 | insol. | sol. in aromatic solvents and chlorinated hydrocarbons |
| Triphenyltin hydroxide | solid | 367 | 118−124 | insol. | sol. in benzene, methanol and other common solvents |

Fig. 102 – This photograph shows a raft trial and the absence of fouling on the tributyltin compound-treated panels can be clearly seen.

Tributyltin benzoate, $(C_4H_9)_3Sn\ OOC.C_6H_5$
Tributyltin benzoate is a clear liquid which can be prepared from tributyltin chloride and potassium benzoate.

It is used as a germicide, usually in combination with a bactericide such as formaldehyde or a quaternary ammonium salt. Aerosol sprays of the disinfectant are available and can be used in hospitals or in the home.

Triphenyltin compounds are gaining importance in crop protection. Fungal diseases cause blight in potatoes, leaf spot and powdery mildew in sugar beets, and attack celeriac, coffee, cocoa, ground nuts, bananas and rice. Triphenyltin compounds have been used successfully against these diseases and are more and more replacing those fungicides which have a tendency to inhibit growth of treated crops. In many cases the organotin is mixed with another pesticide such as a dithiocarbamate. In Europe triphenyltin formulations have now obtained official recognition and in the USA the use of triphenyltin fungicides on potatoes has been approved by the Food and Drug Administration.

Triphenyltin acetate, $(C_6H_5)_3SnOOC.CH_3$
Triphenyltin acetate is a white, crystalline powder prepared by reaction of triphenyltin hydroxide with acetic acid at 80 °C.

The acetate has proved to be an excellent agent against *Phytophora infestans,* potato blight; sugar beets and celery can be protected against *Cercospora beticola* and *Septoria aii* respectively. Commercial products based on triphenyltin acetate are available for this purpose.

Triphenyltin chloride, $(C_6H_5)_3SnCl$
Triphenyltin chloride is a white, crystalline powder prepared by heating tetraphenyltin and stannic chloride or by reacting these chemicals at lower temperatures in the presence of a catalyst.

Triphenyltin chloride finds use in anti-fouling paint formulations and as a preservative for marine timbers, being particularly toxic to marine wood borers such as Teredo worms. The organotin has been tested with promising results as a molluscicide for combating the spread of bilharzia and is an efficient bactericide.

Triphenyltin hydroxide, $(C_6H_5)_3SnOH$
Triphenyltin hydroxide is a white powder prepared from the corresponding chloride by alkaline hydrolysis.

The applications of triphenyltin hydroxide are essentially similar to those of the acetate, viz. as agricultural pesticides and as disinfectants. A marked anti-feedant effect has been noted with both these compounds against plant-feeding larvae and they have been recommended for this purpose.

Two other commercially important compounds are tricyclohexyltin hydroxide and bis(trineophyltin) oxide.

Tricyclohexyltin hydroxide, $(C_6H_{11})_3SnOH$

Tricyclohexyltin hydroxide is a white, nearly odourless powder prepared by alkaline hydrolysis of the chloride. The material is stated to be rather difficult to prepare commercially.

This compound has been shown to be very effective in controlling plant-feeding mites which constitute a nuisance to greenhousemen, nurserymen and growers of deciduous fruits. A commercial formulation based on this organotin has been developed. It is widely used in Europe and has been accepted by the Ministry of Agriculture for use on apple and pear trees in the UK.

Bis(trineophyltin) oxide, bis[tri-(2,2-dimethyl-2-phenylethyl)tin] oxide

This compound is a white crystalline solid, and is used as a specific acaricide for use on apples, pears, peaches and citrus fruits. A commercial formulation based on this organotin has been developed and has been found to give high levels of control against mites which are resistant to organophosphprus and chlorinated hydrocarbon pesticides. It is also claimed to be non-phytotoxic.

### Diorganotin compounds $R_2SnX_2$

The diorganotins correspond to the formula $R_2SnX_2$ where R is an organic group.

The diorganotin compounds are prepared via their halides which are normally obtained from the corresponding $R_4Sn$ compound by reaction with stannic chloride. However there is one important direct route; in this method, tin metal and alkyl halides are reacted at elevated temperatures (around 150 °C) to form dialkyltin dihalides. No solvents are required but a catalyst is necessary. Alkyl iodides and bromides are now being used commercially for this process.

Progressive substitution of the alkyl and aryl groups on tin by other radicals leads to an increase in chemical reactivity. The diorganotin compounds are in general more reactive chemically than the triorganotin compounds although their biological activity is much lower than that of the triorganotins. For this reason the applications of this group of compounds depend largely on the chemical properties.

It is interesting to note that the uses of diorganotin compounds are associated very closely with the plastics industry. The major commercial uses of diorganotin compounds are as stabilisers in polyvinyl chloride (PVC) and in catalytic applications in the production of polyurethane foam and in cold-curing silicone elastomers.

Polyvinyl chloride has a tendency to degrade on heating or on prolonged exposure to light. This degradation results in a yellowing of the plastic and frequently leads to embrittlement. Since fabricating conditions for PVC often involve long periods at relatively high temperatures and, since the product is exposed to light during its life, some means of preventing this degradation is required. It was found that certain chemicals when added to the plastic before processing could inhibit this breakdown and moreover protect it during its service life. Some of the most effective PVC stabilisers known are the dialkyltin compounds, in particular sulphur-containing compounds, which are unsurpassed in conferring heat resistance to the plastic.

The discovery that the dioctyltin compounds are non-toxic as well as being powerful stabilisers was a significant development in opening new markets for PVC. For the first time a powerful heat stabiliser was available which could safely be used in contact with food products. There is now official acceptance in the USA and in many European countries of the use of dioctyltin stabilisers in food packaging grades of PVC. Thus, PVC bottles are used for fruit squashes, vegetable cooking oils, and mineral water. Furthermore, in view of the increasing emphasis being placed on sales appeal in packaged goods which calls for clarity and non-cracking in PVC, these organotins are often used in PVC for packaging toiletries, confectionery, etc. Organotin stabilisers provide maximum clarity under the rigorous processing conditions encountered during the production of PVC bottles and sheets.

A growing market is the use of PVC cladding in the building industry, as shown in Fig. 103, and of clear PVC sheet for roofing applications, where good weatherability and light resistance may be obtained by the presence of certain diorganotin compounds.

Water- and gas-pipe fittings are often made in PVC and use of an organotin stabiliser ensures that high quality extruded parts can be produced even for potable water pipes.

Certain dialkyltin compounds are used as catalysts in the production of polyurethane foams. These foams were originally made in a two-stage process in which, in the first step, a long-chain diol was reacted with an aromatic di-isocyanate, to form a polymer with urethane linkages. In the second stage water was added to the system and reacted with terminal isocyanate groups to form large amounts of carbon dioxide. This gas, in trying to escape from the already thickening polymer, produces a closed-cell honeycomb structure or foam. It was found that certain organotin compounds were not only highly efficient catalysts but caused the first stage to proceed at a much faster rate than the second. This meant that a one-step process could be developed in place of the two-stage method formerly employed.

Dialkyltin catalysts are used in many flexible and rigid polyurethane foams and growth potential for these plastics is good, especially in the furniture and construction industries.

The other major catalytic use of the diorganotins is in room-temperature-vulcanising (RTV) silicones. Under the influence of the tin catalyst, these cure at room temperature from free-flowing liquids or pastes to flexible, elastomeric solids. Applications are wide and range from mould materials for casting plastics or low-melting alloys, to sealants in building and other industries.

Fig. 103 — Cladding and piping made from organotin stabilised PVC are widely
used by the housing industry.

The principal physical and chemical properties of diorganotin compounds are given in Table 28, with additional information in the following pages.

Dimethyltin dichloride, $(CH_3)_2SnCl_2$

Dimethyltin dichloride is a solid at room temperature. The crystals are colourless and the compound is prepared by the reaction of methyl chloride with tin.

Dimethyltin dichloride has been suggested as an alternative to stannic chloride in glass-strengthening treatments. It is claimed that there are no toxic or corrosive breakdown products associated with its use and, if necessary, unused dimethyltin dichloride can be recovered. The technique is essentially similar to that described under tin(IV) chloride.

Dibutyltin dilaurate, $(C_4H_9)_2Sn(OOC.C_{11}H_{23})_2$

Dibutyltin dilaurate is a liquid or low-melting solid, depending on the type and purity of the lauric acid used in its preparation. It is made by reacting dibutyltin oxide with lauric acid.

Dibutyltin dilaurate was one of the first organotin stabilisers to be used for PVC and is still used for lower-temperature processing conditions. The dilaurate is widely used as a catalyst for polyurethane foams and in cold-curing silicone elastomers. Another important use of dibutyltin dilaurate is in treating chickens to cure them of intestinal worm infections; one treatment with 100 mg of the organotin is said to be sufficient.

Dibutyltin maleate, $(C_4H_9)_2Sn(C_4H_2O_4)$

Dibutyltin maleate is a white powder. Depending on the preparation technique, a variety of polymeric structures can be obtained. It can be made by reacting dibutyltin oxide with maleic anhydride in methanol at 60 to 80 °C. It melts at 103 to 105 °C and is insoluble in almost all solvents.

Dibutyltin maleate, like the laurate, was adopted in early processes as a PVC stabiliser. It is still used in spite of some handling problems, such as a tendency to release acrid fumes or to cause stickiness, since it is one of the best light stabilisers known. A modified maleate, the half ester, has been developed and to a large extent overcomes these handling difficulties. PVC for use outdoors, for cladding buildings, often includes the maleate in the stabilising system.

Di(n-octyl)tin maleate polymer

Di(n-octyl)tin maleate polymer is a solid and is an extremely efficient heat stabiliser for PVC, particularly for bottles. This compound has been approved by the US Food and Drug Administration (FDA) for use in PVC for packaging foodstuffs. This approval followed extensive clinical testing of dioctyltins, which showed their mammalian toxicity to be negligible. The authorities of most European countries have also accepted the use of this material in PVC for food contact applications.

**Table 28**

Physical and chemical properties of some diorganotin compounds

| Compound | Physical form | Specific gravity at 20 °C | Molecular weight | Melting point (°C) | Boiling point (°C) | Solubility in | |
|---|---|---|---|---|---|---|---|
| | | | | | | Water | Solvents |
| Dimethyltin dichloride | solid | | 220 | 106–108 | 185–190 | sol. | sol. |
| Dibutyltin diacetate | liquid | 1.32 | 351 | 8.5–10 | 142–145(10 mm) | insol. | sol. organic solvents |
| Dibutyltin di(2-ethyl hexoate) | solid | 1.07 | 519 | 54–60 | 215–220(2 mm) | | |
| Dibutyltin dilaurate | liquid | 1.05 | 632 | 22–27 | | insol. | sol. benzene and acetone |

Di($n$-octyl)tin-$S,S'$-bis (iso-octylmercaptoacetate)
This compound has also been granted FDA approval and the use of this dioctyltin stabiliser has also been sanctioned in most European countries including the UK. It has the advantage of being a liquid, which aids its dispersion in the PVC polymer.

The sulphur-containing diorganotins are the most efficient heat stabilisers known for PVC and give a clear, non-brittle product, even after rigorous processing.

Estertins

These are a relatively new range of diorganotins developed commercially as PVC stabilisers where the R group is an ester rather than, for example, a butyl or methyl group. They offer a range of properties depending on the particular compound and formulation.


**Monoalkyltin compounds $RSnX_3$**
Although the monoalkyl organotin compounds are known they have, so far, only found restricted application. Monobutyltin sulphide, $(C_4H_9SnS_{1\frac{1}{2}})_4$, developed in Germany, is used as a non-toxic stabiliser in PVC film. This application is said to be limited to certain grades of PVC.

Recent research has shown that some of these compounds confer water repelling properties on textiles and masonry (Fig. 104).

Many formulations for stabilisation of PVC contain both mono- and diorganotin compounds which together reinforce the resistance of the material to thermal and ultra-violet degradation.


**FURTHER READING**

1.  Gmelin, L., *Handbuch der Anorganischen Chemie,* Springer-Verlag, Berlin, Vols. 1–6 (Inorganic compounds).
2.  Fuller, M. J., Industrial uses of inorganic tin chemicals, *Tin and its Uses,* No. 103, p. 3, 1975.
3.  Evans, C. J., Developments in the organotin industry. Part 1: Tri-organotin chemicals, *Tin and its Uses,* No. 100, p. 3, 1974; Part 2: Di-organotin compounds, *Tin and its Uses,* No. 101, p. 12, 1974.
4.  Smith, P. J. and Smith, L., Organotin compounds and their applications, *Chemistry in Britain,* Vol. 11, p. 208, 1975.
5.  *Tin Chemicals for Industry,* International Tin Research Institute Publication No. 447.
6.  Mantell, C. L., *Tin*, Hafner Publishing Company, 1970.
7.  Price, J. W., Inorganic tin compounds, *Proceedings of the Conference on Tin Consumption,* International Tin Council, London, p. 201, 1972.

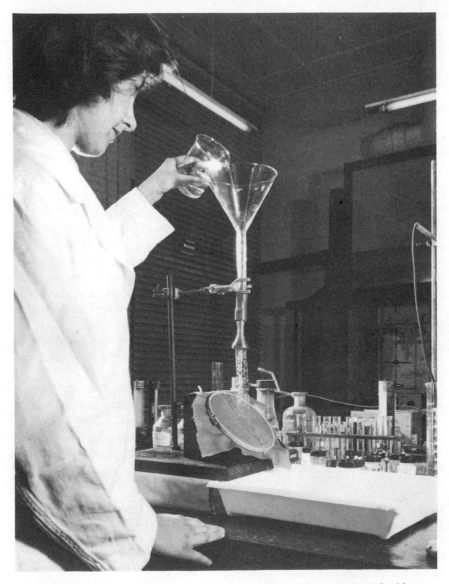

Fig. 104 – Laboratory test for the water-repellency of cotton treated with a solution of a mono-organotin compound. It has also been shown that some mono-organotins confer water repellency on silicates, including bricks and natural minerals.

# 11

# Lesser-known Applications

The principal uses of tin described in the foregoing chapters are for the most part based on knowledge accumulated over many years. There is, however, a definite relationship between the traditional uses of tin and its physical and chemical properties, even though these may not have been fully understood when tin was first adopted for a given use. It is because of the better understanding of the metal's unique properties that the established uses have been continually refined and improved so that they still have a place in modern industrial technology.

It should be noted, however, that in the more recently introduced applications of tin the pattern of development is reversed; new uses have followed from studies of the properties and from research into their exploitation.

This is the pattern which applies generally to the applications described in this chapter. The examples have been selected in order to illustrate the diversity of tin's usage and no claim is made for a comprehensive compilation of the applications of tin, even when taken in conjunction with the preceding chapters. In any case no list could be comprehensive, since new applications are constantly being proposed, tested and adopted.

### Molten tin

The use of molten tin as a means of applying a tin coating has been described in Chapter 8 and its use in the glass industry in Chapter 4. However, there are other processes which exploit the properties of tin in its molten state in metallurgy and in the harnessing of atomic energy.

A completely new use for molten tin is proposed in a new concept for a nuclear reactor now under intensive research in the USA. It relies on the ability of molten tin at high temperature to dissolve nitrogen and thus make it available as a chemical reactant.

The Actinide–Nitride Fuelled (ANF) reactor concept began to be developed in the late 1960s by two researchers in the USA. In their design, molten tin is held under slight nitrogen pressure in a graphite-lined reactor vessel; this means that a small but adequate proportion of nitrogen is in solution in the tin. Uranium is also dissolved in the tin and a certain proportion of this reacts with the dissolved nitrogen to form the nitride of uranium, UN. The uranium nitride is denser than molten tin and so will sink to the bottom of the vessel and when its mass becomes critical, nuclear fission is initiated. If the temperature of the reaction subsequently rises excessively, the uranium nitride will begin to dissociate and uranium goes back into solution, thus diminishing the amount of nitride and slowing down the reaction — an equilibrium which prevents reactions from becoming out of control. Similarly if the temperature drops, the rate of nitriding will increase together with the mass of uranium nitride.

Tin would be a particularly effective solvent metal for this purpose for a number of reasons. Firstly, liquid tin substantially reduces the chemical activity of dissolved uranium due to the tendency of the system to form stable inter-metallic compounds which compete for the uranium in solution. This results in the lowering of the chemical activity of uranium and promotes the reversibility of the nitriding–denitriding reaction. Secondly, since tin has a low melting point and high boiling point, a very wide temperature range over which nitride pre-cipitation can be carried out is permitted. Normally the temperature would be held in the range 1475 to 1800 °C to ensure that the uranium nitride, UN, is precipitated in preference to $U_2N_3$ which precipitates much more slowly. Also, studies of the nitriding behaviour of certain lanthanide and transition metal elements dissolved in tin indicate that those nitride-forming elements present in fission products would not seriously co-precipitate with uranium nitride. The fission product elements would either form low density nitrides that would float to the surface of the tin and be removed or, alternatively, if their concentrations in the solvent tin were low, then their chemical activities would be sufficiently low to allow them to remain in solution with other non-nitride-forming elements.

One of the major possibilities of the ANF reactor concept is that of a recycling process for irradiated nuclear fuels. This method would selectively precipitate only the reactive metals present in higher concentrations; the relatively dense actinide nitrides, including those of uranium and plutonium, would sink and be reclaimed, while the relatively low density fission product nitrides would float to the surface of the molten tin.

Although this process for producing nuclear energy is far from a commercial reality, it represents a radical and interesting new approach to the development of nuclear power generation and a striking example of the exploitation of some of the unique properties of tin.

### Titanium–tin alloys
The increasingly demanding conditions of stress- and temperature-resistance imposed by high performance aero engines and airframes are stimulating a search for new materials capable of withstanding the demands.

Alloys of titanium with aluminium, due to their light weight, high strength and good corrosion resistance, are being increasingly used. Additions of tin to titanium–aluminium alloys have been shown to increase their strength and creep properties without impairing their ductility or the ability to accept hot working.

One of the first titanium–tin alloys to be developed was titanium–5 per cent aluminium–2.5 per cent tin. This is a weldable alloy with good strength at room temperature, which is retained up to moderately high temperatures. This alloy requires hot-forming techniques for fabrication. It has been used for military aircraft construction and for some parts of the service and lunar modules of spacecraft.

In recent years many more complex titanium alloys have been, and are being, developed, especially for high temperature service. A typical example is a complex titanium alloy containing 11 per cent tin, together with aluminium, molybdenum and silicon, which has very high strength and good creep properties up to 450 °C. It is used for various structural components in aircraft (see Fig. 105) and for some of the components in the engines, including those used in Concorde.

### Zircaloys

Zirconium–tin alloys, notably Zircaloy 2 and Zircaloy 4, each containing 1.5 per cent tin, are widely used for cladding the fuel elements in thermal nuclear reactors, on account of their low absorption cross-section for thermal neutrons. good corrosion resistance at elevated temperatures, good mechanical strength and high thermal conductivity.

These zirconium alloys have been developed over the past few decades as part of the programme of research into materials for use in pressurised-water and boiling-water nuclear reactors: see, for example, in Fig. 106.

Early work carried out in the 1950s on corrosion resistance indicated that unalloyed zirconium gave inconsistent results, but that by adding a small proportion of tin and other elements in minor amounts consistency of performance as well as improved corrosion resistance could be attained. This work led first to the development of the alloy Zircaloy 2, which contains 1.5 per cent tin–0.12 per cent iron–0.1 per cent chromium–0.05 per cent nickel, the balance being zirconium.

Further investigation showed that it was possible to reduce the hydrogen pick-up rate of the metal by altering the alloy composition slightly, chiefly by eliminating the nickel and increasing the iron content. Zircaloy 4, which is an alloy containing 1.5 per cent tin–0.21 per cent iron–0.1 per cent chromium–balance zirconium, is chiefly used in the manufacture of fuel rod cladding and for pressure tubes and structural parts for pressurised water reactors.

### Tin in high temperature coatings

Tin is also finding a new use in the technology of high-temperature materials as a constituent of surface coatings for refractory metals. To improve the efficiency

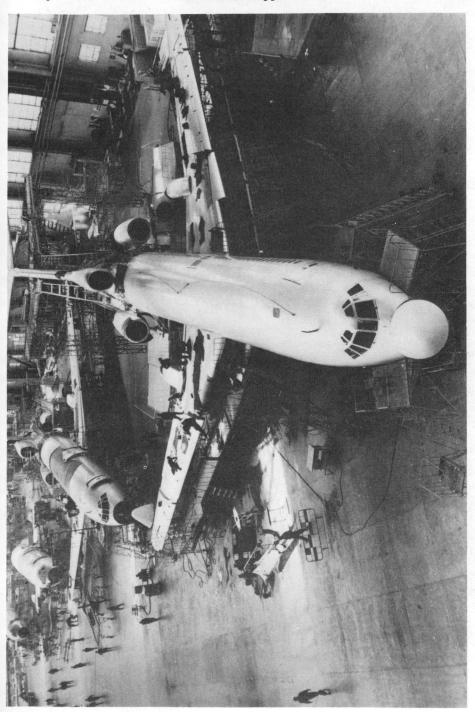

Fig. 105 — Tin-containing titanium alloys are widely used in aero-engines and airframes. (Courtesy: Novosti Press Agency.)

Fig. 106 − Zirconium−tin alloy clad fuel element being inserted in the Beznau
(Switzerland) nuclear reactor. (Courtesy: Westinghouse Electric S.A.)

of jet turbine engines, there is a need for materials which will operate at temperatures well above 1000 °C. High melting point metals such as tantalum and molybdenum and their alloys have the desired creep properties at high temperatures but their poor resistance to oxidation renders them useless unless the surface can be protected.

A suitable coating must not only afford protection, it must also continue to do so under repeated conditions of heating and cooling. A coating developed for this purpose is obtained by applying a mixture of tin and aluminium powders to the surface to be protected. During subsequent heat treatment, the aluminium reacts with the basis metal to form an intermetallic compound. Although tin and aluminium have no mutual solubility in the solid state, at the high operating temperature in, for example, a jet engine, the tin acts as a liquid carrier for the aluminium, thereby healing cracks in the coating which may develop during the heating/cooling cycle.

Various combinations of tin plus aluminium as self-healing surface protection have been successfully developed such as, for example, a mixture containing equal parts of tin and aluminium used to form a protective coating for tantalum sheet intended for the combustion chambers of ram-jet engines.

### Superconducting alloys

As discussed in Chapter 1, an important characteristic of tin is its ability to form intermetallic compounds with many other metals. Some of the intermetallic compounds of tin have unique properties and hence are finding important scientific and technological outlets. An example of this is in the field of superconductors.

The phenomenon of superconductivity was discovered early in the present century (by Kamerlingh Onnes) when it was found that at very low temperatures, approaching absolute zero (0 K, $-273.25$ °C), the electrical resistivity of certain metals is reduced virtually to zero or to a very low value, and the metal is said to be superconducting. Metallic tin, for example, becomes a superconductor at about 4 K ($-269$ °C).

Since it is the electrical resistivity of a metal that limits its current-carrying ability, a metal in the superconducting state is obviously of enormous potential interest both for the production of large magnetic fields and as a conductor of high electrical currents for transmission over long distances with negligible power losses. However, to make practical use of superconductivity, it is necessary to develop materials which will remain superconductive at temperatures appreciably above absolute zero (say around 15 K, $-258$ °C) and in strong magnetic fields.

One of the most successful superconductors that has so far been developed, because it has one of the highest superconducting transition temperatures, is the intermetallic niobium—tin compound $Nb_3Sn$. This retains some of its superconductivity at magnetic fields up to 200 000 gauss. It has been used in the construction of powerful electromagnets in which power consumption is almost zero; liquid helium (b.p. 4.55 K, −268 °C) is used to maintain the superconductor at its operating temperature.

One of the problems originally associated with the use of the niobium—tin material was in fabricating it. The intermetallic compound $Nb_3Sn$ is brittle and hence difficult to draw into the wire necessary to wind the coils of the electromagnet, but alternative processes have been developed. One is to produce a very thin film of $Nb_3Sn$ on a more ductile base; another involves filling a niobium tube with tin, drawing this down to wire, winding the coil and finally heat treating the coil so that the tin reacts with the niobium, forming $Nb_3Sn$ at the interface. A third procedure is to embed filaments of niobium in a drawn tin-bronze rod and heat treat so that the niobium is converted to $Nb_3Sn$ by withdrawal of tin out of the bronze.

At present superconducting electromagnets are used chiefly for research with a view to developing commercial electrical power generators and also for research in nuclear physics, where strong magnetic fields are required, as illustrated in Fig. 107.

## Semiconductors

A further example of an application for a tin intermetallic compound is of recent origin and results from the exploitation of yet another property of a tin-containing material. As has been pointed out in earlier chapters, the basic characteristics of tin arising from its atomic structure enable it to exhibit either predominantly metallic properties or non-metallic properties, depending on its immediate environment. Therefore, in its metallurgical and inorganic chemical applications tin behaves like a metal, whereas in organotin compounds tin behaves chemically like a carbon atom.

In recent years there has been considerable development in the field of solid state devices for electronic applications and a consequent search for materials with a wide range of semiconductive properties. Whilst in many applications transistors make use of semiconductors with a relatively high energy gap of around 1 eV, for work in conjunction with lasers the need is for a semiconductor with a small energy gap.

The intermetallic compound SnTe between tin and tellurium, which behaves more like a chemical compound than a metal, has a rock salt crystal structure and an energy gap of about 0.3 eV at room temperature. This compound forms a complete series of 'solid solutions' with the corresponding lead—tellurium compound, forming a pseudo-binary system. This means that by varying the tin—lead ratio it is possible to form compounds of the composition $TeSn_xPb_{(1-x)}$.

This intermetallic compound can be used to emit, or to detect, laser light at specific wavelengths in the infrared range. By selecting suitable wavelengths (and hence compositions) characteristic of different gas molecules, the concentration of specific gases in the atmosphere can be determined. It is estimated that such a

Fig. 107 – Electromagnet with $Nb_3Sn$ windings used in neutron diffraction
equipment. (Courtesy: Oxford Instruments Co. Ltd.)

semiconductor laser can easily detect pollution in the air by a gas whose 'concentration' is as low as one part in a thousand million. Tin—tellurium semiconductors are coming into increasing use in a variety of other applications.

**Tin in dental alloys**

The uses of tin are so diverse that one or other must beneficially affect the lives of everyone, even if in an unsuspected way. It is perhaps fitting, then, to conclude this survey of the lesser-known uses of tin with one which is common to most of us, even if unfamiliar.

The use of tin in dentistry is one of long standing and stems in part from its non-toxicity. In mediaeval times pure tin foil was used for filling holes in teeth, but over the years there has been a progressive improvement in the materials employed. In modern dentistry the need is for a substance which can be packed into a cavity under pressure whilst plastic and which will subsequently set hard at ambient temperature without any shrinkage. The compound employed today for the majority of dental fillings is the tin—silver intermetallic, $Ag_3Sn$, which is the basis for dental amalgams.

The compound normally used is made by melting silver with approximately 25 per cent tin, plus small additions of copper and zinc. The compound $Ag_3Sn$ which is formed is sufficiently brittle to be crushed to powder and this powder is subsequently amalgamated with mercury by the dentist immediately prior to insertion, in order to obtain a plastic material. The components of modern dental amalgams are pre-packed in special plastic vials which enable amalgamation to be carried out in the sealed vessel. The process of setting is similar to liquid phase sintering and a mercury—tin compound is formed. Whilst this use of tin does not represent a major tonnage consumption, it is, nevertheless, of very wide application!

**FURTHER READING**

1. Dinsdale, P. M. and Evans, C. J., High performance alloys, *Tin and its Uses*, No. 119, p. 1, 1979.
2. Long, J. B. and Evans, C. J., Tin-containing brazing alloys, *Tin and its Uses*, No. 118, p. 13, 1978.
3. Muller, B. F., Tin for wine bottle closures, *Tin and its Uses*, No. 114, p. 15, 1977.
4. Dinsdale, P. M., Lead—calcium—tin alloys for the lead—acid battery, *Tin and its Uses*, No. 110, p. 12, 1976 and No. 111, p. 13, 1977.
5. Niobium—tin superconductors for industry, *Tin and its Uses*, No. 109, p. 11, 1976.
6. Conductive tin oxide films on glass by vapour deposition, *Tin and its Uses*, No. 107, p. 10, 1976.
7. Robins, D. A., Some lesser known uses of tin, *Proceedings of the Conference on Tin Consumption,* International Tin Council, London, p. 269, 1972.

# 12

# Economic Factors

---

## 12.1 DISTRIBUTION OF PRODUCTION AND USE

In several important respects, tin differs markedly from the other major base metals. In the first place, although tin is an important industrial metal, its total world production at around 200 000 tonnes annually is small when compared with the other common non-ferrous metals. Most of these measure their production in millions of tonnes (Table 29), the principal exception being nickel, and even this metal has an annual production of about three times that of tin.

### Table 29

World primary production of common non-ferrous metals
(annual average 1970–1980)

| Metal | Annual production (millions of tonnes) |
|---|---|
| Aluminium | 13.3 |
| Copper | 7.3 |
| Zinc | 5.8 |
| Lead | 3.4 |
| Nickel | 0.67 |
| Tin | 0.19 |

*Source:* Metal Statistics – Metallgesellschaft AG.

In the second place, major tin deposits are confined to a comparatively small number of areas (Fig. 108). More than half the world's tin ore is mined in the great Far Eastern tin field which stretches from Burma and runs down through Thailand and Malaysia to the islands of Indonesia. Bolivia is a major producer and for many years ranked second in the league of tin-producing countries. The only

Fig. 108 — Map illustrating location of the major tin mining regions of the world.
1: Bolivia. 2: Brazil. 3: Nigeria. 4: Zaire. 5: South Africa. 6. USSR. 7: China
(People's Republic). 8: Thailand. 9: Malaysia. 10: Indonesia. 11: Australia. 12: UK.

other area of significance in the American continent is Brazil, where tin deposits
of economic importance have been found and begun to be exploited in recent
years; otherwise no major workable deposits are known either in North or South
America, despite the wealth of other minerals in the continent. In Africa, tin is
found in significant amounts only in Nigeria, Zaire and South Africa. Australia is
a major producer from a number of small tin fields, chiefly located in Tasmania,
Queensland and New South Wales.

Unlike the other major tin producers, Australia is also a substantial consumer of tin, and domestic consumption accounts for almost half of the mineral
output. Although centuries ago European tin deposits were an important source
of tin, with recorded outputs from England, Spain, Saxony and Bohemia, today
there is very little tin mining practised in Europe, the principal exception being
in England where the Cornish tin deposits continue to be worked, even after
many centuries of exploitation dating back to at least Roman times. The major
tin-producing countries are listed in Table 30 which also gives the tin consumption of the major tin ore producing countries. This highlights the fact that the
four main producers use negligible quantities of tn. The reason is, of course, that
the applications of tin are mainly in high technology industries and these have
yet to be established in these producing countries. Conversely, those countries

**Table 30**

Major sources of tin ore

| Country[1] | Production of tin-in-concentrates[2] (tonnes) | Consumption of tin metal (tonnes) |
|---|---|---|
| Malaysia | 66 000 | 300 |
| Bolivia | 30 300 | 380 |
| Indonesia | 26 300 | 400 |
| Thailand | 25 000 | 420 |
| Australia | 11 200 | 3 580 |
| Brazil | 5 100 | 3 860 |
| Nigeria | 4 300 | 80 |
| Zaire | 4 200 | 120 |
| United Kingdom | 3 100 | 12 480 |
| South Africa | 2 600 | 2 130 |
| Rwanda | 1 400 | 2 130 |
| Zimbabwe | 1 000 | 2 130 |
| Sub-total | 180 500 | 23 750 |
| Other countries[3] | 11 000 | 162 850 |
| World Total | 191 500 | 186 600 |

*Notes:* (1)  Countries producing in excess of 1000t/annum.

(2)  Average production 1971–1981.

(3)  Includes production from over 20 countries who are minor producers.

which are major consumers of tin, notably USA, Japan and the EEC countries (Table 31) have virtually no workable tin deposits. Here again an exception to this general pattern is the UK, which is a large user who yet can produce up to about 25 per cent of its requirements. Of the net tin producers, Australia and Brazil both have important tinplate industries which use 60 per cent and 40 per cent respectively of the tin consumed, together with large-scale manufacture of electrical and electronic products.

Another important respect in which tin deposits differ from those of other metals is that they are relatively small. There are many copper, zinc or lead mines which produce a hundred or more tonnes of metal a day. There are few tin mines which have produced more than 10 tonnes of metal per day, although there are hundreds of mines which produce less than 20 tonnes of tin in a year.

## Table 31

Major tin consuming countries

| | | Average consumption 1971–1981 (tonnes) |
|---|---|---|
| *Africa* | | |
| | South Africa | 2100 |
| | Other | 1600 |
| | Total Africa | 3700 |
| *N. and S. America* | | |
| | USA | 49500 |
| | Canada | 4700 |
| | Other (incl. Brazil) | 9000 |
| | Total America | 63200 |
| *Asia* | | |
| | Japan | 31700 |
| | Other | 10500 |
| | Total Asia | 42200 |
| *Europe* | | |
| | EEC | 52100 |
| | Other | 21500 |
| | Total Europe | 73600 |
| *Oceania* | | |
| | Australia | 3580 |
| | Other | 320 |
| | Total Oceania | 3900 |
| **World Total** | | **186600** |

*Source:* ITC Tin Statistics 1971–1981.

The very large number of entirely separate and frequently small tin mines, coupled with the wide diversity of uses of tin — again, often in relatively small quantities — in part account for the fact that the tin industry, again unlike most metals industries, is not 'vertically integrated'; that is to say, there is normally no financial link between the miner, the smelter and the intermediate or ultimate consumers.

One major change which has taken place in the production of tin over the past 20 years or so has been the geographical shift of tin smelting to the countries in which the ore is mined. Formerly the major ore producing countries shipped the bulk of their product as concentrates (see Chapter 2) to be smelted in the industrial countries, principally in Europe, where Britain, the Netherlands and Belgium were all important primary tin smelting countries. Today, there is a growing trend to have tin concentrates smelted in their country of origin. A result has been that most European smelters have been closed or have turned to treating recycled or secondary material.

Tin was smelted in Malaysia as early as 1902, but in the other major producing countries smelters were built more recently, for example in 1962 in Nigeria, in 1965 in Thailand, in 1967 in Indonesia, and in 1970 in Bolivia. The Australian smelter dates from 1967 and tin is also smelted in Brazil. The majority of these produce only tin metal and do not engage in the production of tin alloys such as solder. Table 32 gives the world production figures for tin-in-concentrate (i.e. as mined) and for primary tin metal as produced by the smelters. The same table also gives the average annual price of standard tin on the London Metal Exchange (LME) over the same period (1971–1981). The high price of tin, as compared with the other non-ferrous metals, is in part accounted for by the leanness of the ores and by the wide dissemination of the deposits. The tin content of the ore mined ranges from about 1 per cent in primary lode mines to as little as 0.015 per cent in alluvial deposits. This is seen in perspective when

**Table 32**

World tin production, consumption price

| Year | Tin-in-concentrates production ('000 tonnes) | Primary tin metal | | Average Annual LME price (£/tonne) |
|------|------|------|------|------|
| | | Production ('000 tonnes) | Consumption ('000 tonnes) | |
| 1971 | 188.1 | 186.5 | 188.6 | 1437 |
| 1972 | 196.3 | 191.4 | 191.7 | 1506 |
| 1973 | 189.1 | 187.8 | 214.1 | 1960 |
| 1974 | 183.6 | 181.5 | 200.1 | 3494 |
| 1975 | 181.2 | 178.5 | 173.8 | 3091 |
| 1976 | 180.0 | 182.5 | 193.5 | 4255 |
| 1977 | 188.4 | 179.9 | 184.4 | 6181 |
| 1978 | 196.7 | 193.5 | 184.6 | 6706 |
| 1979 | 200.4 | 201.3 | 185.6 | 7276 |
| 1980 | 200.8 | 197.9 | 174.2 | 7222 |
| 1981 | 201.9 | 195.5 | 162.1 | 7085 |

*Source:* ITC Tin Statistics 1971–1981.

one considers that ores of lead, for example, are rarely less than 2 per cent, while an iron deposit of less than 50 per cent may be considered of little importance.

However, as will be noted from the uses of tin described in earlier chapters, tin is rarely used as the major constituent of an alloy, or of a component, so that in many instances, the price of tin is not the governing factor in its use. Nevertheless, because it is intrinsically valuable, tin is under constant threat of substitution, the degree of which varies with the application. In tinplate, for example, the tin may represent only 0.2% of the mass of the container, but for certain packaging applications, such as beer cans, there is strong competition from alternative materials, including glass and plastics. Conversely, in electronics applications, the cost of the solder relative to the complete assembly (e.g. television receiver) is not significant.

## 12.2 THE INTERNATIONAL TIN AGREEMENTS

As has been suggested above, most of the countries which are the major consumers of tin have little or no available tin deposits. Conversely, the major tin producers are relatively small consumers. Furthermore, the national economies of the developing country producers are dependent on tin exports as a major source of their revenue. This means that tin is an extremely important commodity in terms of international trade, and it is one of the few commodities — and currently the only metal — whose trade is controlled by an International Agreement, set up under the auspices of the United Nations, and comprising both producer and consumer governments.

The International Tin Agreement is implemented by an inter-governmental body consisting of representatives of both producing and consuming countries, namely the International Tin Council (ITC). The Council was set up in 1956 and since then has operated six successive International Tin Agreements whose objectives have been, *inter alia*, to achieve a long-term balance between world production and consumption of tin, to alleviate serious difficulties arising from an actual or anticipated surplus or shortage, and to prevent excessive fluctuations in the price of, and earnings from, tin. The Council is made up of representatives of the governments of producing and consuming countries, and meets in full session at least four times a year to examine the economics of the tin industry and the market. In any matter decided by voting, the consumer members' total number of votes is equal to that of the producers'. This means that, contrary to popular opinion, the ITC is not a producer 'cartel', but a forum for genuine consultation and agreement between consumers and producers.

Among its objectives, the Tin Agreement, through the ITC, seeks to balance supply and demand by means of a buffer stock and, if necessary — in certain defined circumstances of serious surplus or decreased demand — by export controls. The buffer stock's sales or purchases of tin are made on the established metal markets. The Buffer Stock Manager operates within price ranges determined at Council sessions, where producing and consuming countries have an equal number of votes, and where approval of the majority of both producers and consumers is required to bring about any change; consequently the ITC price range is not comparable with a 'producer price' as is sometimes supposed. The buffer stock operation seeks to reduce fluctuations, while allowing the forces of supply and demand to determine the long-term price trend.

In determining the price range, the Council takes into account the short-term developments and medium-term trends of tin production and consumption, the existing capacity for mine production, the adequacy of the current price to maintain sufficient future mine production capacity, and other relevant factors affecting movements in the tin price, such as the economics of tin-consuming industries.

Between the floor and ceiling prices, three sectors are established within which the Buffer Stock Manager operates, subject to certain limitations imposed upon him by the Council. The Manager must sell tin if the market price is equal to, or greater than, the ceiling price until either the price has fallen below the ceiling or until the stocks of tin at his disposal are exhausted. He may sell or buy tin if the price is in the upper sector, if he considers it necessary to prevent the market price from rising too steeply, but he must remain a net seller. The converse applies in the lower sector, the Manager is obliged to buy tin when the price is equal to, or below, the floor price, until either the price rises above the floor or the funds at his disposal are exhausted.

## 12.3  TRADING IN TIN

There are two key markets at present which exert an important influence on the international trade in tin. These are the Penang Market in Malaysia, which is a centre for physical selling of tin, and the London Metal Exchange, which is a forum for commercial transactions in non-ferrous metals and includes tin amongst its commodities. Tin is also traded on some other major commodity exchanges, for example in New York.

### The Penang Market
In contrast to the general commodity markets, the tin price on the Penang Market is set by the actual sale of tin metal by the Penang smelters to dealers and consumers, the price being determined by bids made for tin to be sold on that day. The amount of tin for sale is governed by the content of tin in the concentrates sold to the smelters on that day.

The sequence of events is as follows: the tin concentrates, representing the product of a number of individual mines, are purchased outright for cash by the smelters, who then estimate the tin content. This represents the day's offering and the smelters confer to decide how much tin they have available for that day's sales. Meanwhile, bids have been submitted in writing to the smelters and these bids are matched against the estimated supplies of tin. The price fixed for the day's trading is the highest single figure at which these tin supplies can be cleared. Each day, about 30 to 40 bids may be received, usually totalling more than the average quantity of 250 tonnes which are typically for sale daily. Those bids which were originally higher than the price finally fixed are then met completely, whilst the remainder of the tin available is shared amongst those whose bids matched the final price.

The Penang price thus determined serves as a general indicator of tin prices and forms a guide for other trading concerns. The producer thus sells his concentrate to the smelter (which means the miner is paid immediately on delivery), the smelter despatches his purchases to the smelting works and subsequently delivers the tin metal to the purchaser within a contracted 60 days.

Thus, although in the first instance, it is tin concentrates which are bought and sold, the net result is that tin metal changes hands.

### The London Metal Exchange

The London Metal Exchange (LME) was established during the nineteenth century for trading purposes, and its present day pre-eminence as a centre for trade in non-ferrous metals can be judged from the fact that quotations on the LME are widely regarded today as the basis for world prices. Dealings in tin take place in 'lots' of 5 tonnes and are for either immediate delivery or for delivery in 3 months; hence the two prices quoted in market reports − cash and forward.

The LME is an 'open outcry' market and the business is conducted in a hall in which the principal traders are seated on a circular, leather-covered bench known as the 'Ring'. All transactions are conducted verbally by the traders who call out the quantity of metal they wish to buy or sell and the price of the bid or offer. Details of negotiated deals are recorded by assistants standing behind the ring and the deals are later confirmed by official contracts.

Tin is one of the seven metals currently traded (the others being copper, lead, zinc, silver, nickel and aluminium) and there are four periods on each market day when official business in the metal is conducted. Two of these periods are in the morning and two in the afternoon. Prices are arrived at as a result of many bids and offers, but, once the settlement prices are established each day, these are taken as the 'closing prices'. Business, however, may continue for a period on the 'kerb' market. This term arises from the early days of the LME when dealings continued on the pavement outside after the Exchange was closed. In practice, unofficial trading continues at all other times of the day.

### 12.4 THE PATTERN OF TIN USAGE

The foregoing chapters in this monograph have described in some detail the most important applications of tin, its alloys and compounds. In this section, it is of interest to note the pattern of consumption, although the relative tonnages used for the different end-uses do not necessarily reflect the industrial importance. This is because, as has been shown, tin is in many instances a minor ingredient by proportion but a vital constituent in terms of the properties it imparts.

The industrial uses of tin are summarised in Table 33. This indicates the wide variety of industries that are influenced beneficially by tin.

The relative proportion of tin used in the major end-use classifications is shown in Table 34.

The pattern of tin usage, like that of most other metals, changes very slowly, using the material vary as technology develops. For example, the newer methods of can manufacture mean that less tin is consumed per tonne of tinplate produced, due to the trend towards thinner tin coatings. Overall, however, this is

**Table 33**

Summary of industrial uses of tin

| | | |
|---|---|---|
| Tinplate | Food cans | Open-top and general line containers; pet food and beer and beverage cans. |
| | Other containers | For oils, chemicals, paints, cosmetics, and other non-food products. |
| | Closures | Screw caps, twist-off caps, crown corks, etc. |
| | Engineering and electrical | Automobile, radio and electrical applications; gas meters; general light engineering and press work. |
| | Other uses | Kitchen and dairy equipment; display and advertising signs; toys. |
| Tin and tin alloy coatings | Pure tin | Food processing and transport equipment; water-heating and cooking equipment and utensils; electrical and electronic equipment; tags, eyelets and fasteners; steel and copper wire; pistons and pump parts. |
| | Tin–lead | Electrical and electronic equipment; fire extinguishers; gas equipment and fuel tanks; terneplate. |
| | Bronze | Haberdashery; utensils and ornaments; undercoats for other metal coatings. |
| | Tin–nickel | Instrument and apparatus components; haberdashery; reflectors; printed circuits. |
| | Tin–zinc | Hydraulic equipment; electrical equipment; automobile and motor-cycle components; fire extinguishers. |

**Table 33** – *continued*

| Tin alloys | Tin | Foil; pipes; collapsible tubes; fuses; powder metallurgy. |
|---|---|---|
| | Solder | Electrical, electronic industries; metal containers; plumbing; automobile radiators and heat exchangers; all metal-joining tasks. |
| | Bronze | Marine, chemical and general engineering; bearings and bushes; pumps; musical instruments; architecture; coinage. |
| | Tin-base | Bearings and bushes; costume jewellery; precision castings; organ pipes; pewter and art work. |
| | Fusible alloys | Sprinkler, alarm and safety devices; low-melting solders; metal seals; textile industry; press tools; moulds. |
| | Aluminium–tin | Bearings; pump castings. |
| | Cast iron | Pearlitic, wear-resistant and heat-resistant irons. |
| | Miscellaneous | Printing and die-casting metal; dental alloys; rare-metal alloys. |
| Molten tin | | Float glass; sealants; oxidation-resistance systems. |
| Inorganic compounds | | Catalysts and reductants; opacifiers and colourants; electrodeposition; veterinary medicines; silvering; glass coatings; toothpaste and soap; abrasives. |
| Organic compounds | | Catalysts; stabilisers; fungicides; insecticides; disinfectants; chemical intermediates. |

**Table 34**

World consumption of primary tin by end-use (%)

| | |
|---|---|
| Tinplate | 40 |
| Solder | 28 |
| Whitemetal and pewter alloys | 7 |
| Bronze | 6 |
| Tinning | 4 |
| Chemicals | 7 |
| Miscellaneous | 8 |
| | 100 |

largely compensated for by an expansion of tinplate production and use into more countries. Similarly, the bulk uses of solders in the early years of the century have given way to electronics as the major user of tin solders. The more recent use of microcircuits leads to less solder per component, but again this is more than compensated by the staggering increase in electronic products in use, so that tin consumption is maintained in tonnage terms. The past 20 to 30 years have seen a steady increase in the consumption of tin in chemical compounds. This trend is continuing and the chemical industry represents a major potential growth area.

Whilst the actual consumption of tin may fluctuate on an annual basis, the pattern throughout the present century has been one of a steady, if unspectacular, growth of around 1 to 2 per cent per annum. With no immediate prospects of sudden gains, or of major losses to competitive materials, it would seem likely that this steady growth pattern will continue in the longer term.

## 12.5 THE INTERNATIONAL TIN RESEARCH COUNCIL

The International Tin Research Council was established in 1932 with the object of developing the use of tin. The countries which form the International Tin Research Council are also "producer members" of the International Tin Council, but the two organisations are separate and have different functions.

The executive arm of the International Tin Research Council is the International Tin Research Institute. It is financed entirely by tin producers, but its researches are concerned with the consumption and applications of tin. In fulfilment of its purpose, the Institute carries out research and development work based on a scientific study of the metal and of the processes which use tin, or which could provide future markets. The International Tin Research Council also maintains nine technical information centres in major tin-consuming countries.

In its 50 years of existence, the Institute has discovered and developed many new uses of tin, many of which are described in this monograph.

## FURTHER READING

1.  Allen, H. W., How the Tin Agreement Works, *Proceedings of the Conference on Tin Consumption,* International Tin Council, London, p. 451, 1972.
2.  Fox, W., *Tin: The Working of a Commodity Agreement,* Mining Journal Books, London, 1974.
3.  *Statistical Bulletin,* International Tin Council, London (monthly).
4.  *Metal Statistics,* Metallgesellschaft AG, Frankfurt (annual).
5.  *Technological Developments in Tin Consumption Combat Substitution,* International Tin Research Institute Publication No. 541.
6.  Gibson-Jarvie, R., *The London Metal Exchange,* Metallgesellschaft AG, Frankfurt am Main, 1976.
7.  Robertson, W., *Tin: Its Production and Marketing,* Croom Helm Commodity Series, 1982.

# Appendices

## 1. METALLOGRAPHY

Tin and tin-rich alloys are extremely soft and have a low recrystallisation temperature, so that it is difficult to obtain an unworked surface free from scratches. This may result in the appearance of false surface structures and lead to misinterpretation of metallographic data unless considerable care is taken during polishing and etching to avoid surface distortion.

### Mounting

Tin alloy specimens should be mounted in room temperature-curing materials, because elevated temperatures can cause structural changes such as precipitation of a second phase from a supersaturated solid solution or the formation of intermetallic layers between tin coatings and some substrate metals. Thermo-setting mounting compounds therefore are not suitable. Specimens can be mounted satisfactorily at room temperature, normally without the application of pressure, using materials such as polyesters and acrylics. It is recommended that mounting compounds exhibiting a low heat of reaction should be selected to limit the temperature rise during curing.

When it is desired to examine the edges of a soft tin-rich specimen, or of a tin coating on a harder basis metal, a supporting layer of copper or perhaps nickel is electrodeposited on to the surface before mounting. Before being electroplated, the specimen must be solvent degreased in hot industrial methylated spirit and then cathodically cleaned in a 5 per cent solution of trisodium phosphate. (For tin coated materials, this step can be omitted in order to avoid risk of attack on the coating.) An initial strike coating of copper about 1 $\mu$m thick is deposited from a cyanide bath. The deposit may be subsequently built up to a thickness of about 50 $\mu$m in the same electrolyte or in a low stress, dull nickel bath, depending on the hardness of the specimen. Copper is the more generally used backing-up material because, unless care is taken, the stress in a nickel deposit may cause it to detach from the sample. Some success can be achieved without electroplating a support coating if fine iron, copper or nickel powder is mixed with the mounting plastic to give it a greater effective hardness. However it should be noted that it is difficult to avoid 'scooping out' of tin coatings on harder basis metals if the tin coating is thick (for example more than about 10 $\mu$m) even when a support layer is plated on.

When electroplated or similar very thin coatings are required to be examined, oblique specimens called taper sections are often used. Use of taper sections increases the effective magnification in the direction normal to the interface and so facilitates examination of the coating and of interface reaction products such as intermetallic compound layers. Experience has shown that a true structure is obtained only along the edge adjacent to the acute angle (Fig. 90).

A linear enlargement of 10 is obtained by using taper sections mounted to give an angle between the section plane and the specimen axis of $6°$.

A two-stage mounting process is used to obtain taper sections. The specimen is first mounted, with its coated surface down, on a tapered mounting plug with one face sloped at $6°$ ($84°$ to the vertical axis). After being extracted from the mounting die, the tapered mount holding the specimen is inverted and placed on a straight mounting plug in the die and completely covered with an additional layer of mounting compound. After being allowed to set and fuse with the original tapered mount, this additional mounting compound is removed by machining until the edge of the specimen is just revealed. The specimen is subsequently polished and etched for examination. The taper mounting procedure is illustrated in Fig. 109.

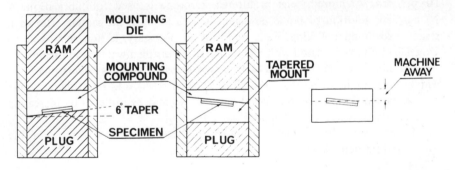

Fig. 109 – Diagram illustrating the two stages of preparing metallographic mounts for taper sections.

### Grinding and polishing

With tin, its alloys and coatings, it is essential to avoid distortion of the surface regions which will result in obtaining false microstructures.

To avoid working the surface regions, extreme care must be taken during the initial machining or filing of mounted specimens as well as in the subsequent grinding and polishing. Fairly light pressure is preferred during all stages of polishing and in each stage polishing should be continued for a period twide as long as that necessary to remove the scratches from the previous stage.

Specimens are flattened on a file or turned in a lathe using progressively smaller cuts to a final series of three or four cuts each of about 15 $\mu$m depth. They are then ground on silicon carbide papers having progressively finer grit sizes of 220, 320, 400 and 600 mesh. The papers are kept wet by a continuous stream of lubricant, usually water, that washes away particles of metal as they are cut from the surface of the specimen. Kerosene is used for materials which might be attacked by water.

Light pressure only is used and the samples are rotated through 90° periodically, usually when transferring to the next finest abrasive paper. Occasionally it is preferred to omit the use of silicon carbide for pure tin and other soft alloys because the abrasive tends to embed. The turned or filed surface would then be ground automatically on two rotating pads, the first impregnated with 60 $\mu$m diamond and the second with 25 $\mu$m diamond, before the polishing stages described below.

The scratches from the grinding are removed by polishing for several minutes on a rotating wheel covered with a short nap or napless cloth impregnated with 6 $\mu$m diamond compound. The specimen is then polished on a succession of wheels covered with microcloth impregnated with 1, 0.25 and 0.1 $\mu$m diamond compound respectively. For best results, the last two stages may be performed by hand polishing rather than on a rotating wheel. A solution consisting of 10 per cent trichlorethylene in filtered kerosene is used for lubrication. A 20 per cent solution of liquid detergent is used to wash the sample after each stage of diamond polishing. Final polishing, if necessary, is done by hand on long nap cloth impregnated with $\gamma$-alumina, using distilled water as lubricant.

To minimise edge-bevelling or interface steps in samples having phases of widely differing hardness (for example tin-coated steel), a fast cutting rate must be maintained during diamond polishing and this may be achieved by increasing the quantity of diamond on the polishing pad.

Electrolytic polishing methods are not particularly suitable or reproducible for tin and tin-rich alloys.

**Etching**
Different etchants are used for revealing the structure of the various alloys depending on where the contrast or detail is required. Table 35 summarises the etchants used and their specific purposes.

Examples of the use of some of these solutions are given in Figs. 110 to 112. Many of the microstructures of the commercially significant tin-containing alloys have been illustrated in Chapter 3; there is therefore no need to repeat them here. However, a further selection of photomicrographs is given in this appendix especially to illustrate the effect of the different etching solutions.

**Table 35**

Etchants for use in microscopic examination of tin and tin alloys

| Etchant composition | Uses |
|---|---|
| 4 g Iron (III) Chloride<br>10 ml Hydrochloric Acid<br>60 ml Water<br>120 ml Industrial Methylated Spirits<br>(IMS) | General use for tin and tin alloys<br>(Grain contrast — Fig. 110) |
| 2 ml HCl, 98 ml | Grain-boundary etch for pure tin |
| 10 ml $HNO_3$, 10 ml acetic acid,<br>80 ml glycerol | Darkens the lead in the eutectic of<br>tin-rich tin–lead alloys |
| 5% silver nitrate in water | Darkens primary and eutectic lead in<br>lead-rich tin–lead alloys (Fig. 111) |
| 2 ml Nitric Acid<br>98 ml Industrial Methylated<br>Spirits (nital) | Recommended for etching tin–antimony<br>alloys; darkens tin-rich matrix, leaving<br>intermetallic compounds unattacked.<br>Often used for etching specimens of<br>babbitted bearings. |
| 5 g Picric Acid<br>95 ml Industrial Methylated<br>Spirits (picral) | For etching tin-coated steel and<br>tin-coated cast iron (see text) |
| 1 drop $HNO_3$ (conc.), 2 drops HF,<br>25 ml glycerol; then picral | For etching tin-coated steel<br>(see text) |
| Ammonium hydroxide (S.G. 0.880)<br>containing a few drops of 20 vol (6%)<br>hydrogen peroxide | For etching tin-coated copper and<br>copper alloys (Fig. 112) |

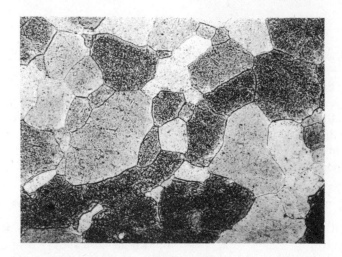

Fig. 110 – Grain contrast etch for pure tin (see Table 24). (Magnification ×250.)

Fig. 111 – Coring in tin, 2.5% antimony alloy revealed by etching (see Table 24). (Magnification ×100.)

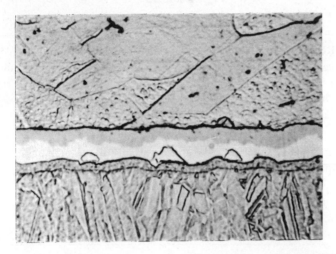

Fig. 112 – Hot dipped tin coating on copper etched to show intermetallic ($Cu_6Sn_5$) layer/substrate interface (see Table 24). Sample copper-plated over the tin coating to prevent bevelling during polishing. (Magnification ×2000.)

## Electron microscopy
### Transmission method

Two standard techniques exist for examination of samples in the transmission electron microscope (TEM), these being the replica and thin-film techniques. The production of thin films of tin and tin-rich alloys by electrochemical dissolution may be classed as difficult. By the newer techniques of ion bombardment some success has been claimed. Alloys containing a small amount of tin, such as tin bronze, may be thinned by slight variation of the condition for thinning the major element (for example copper) but for satisfactory results the tin should preferably be in solid solution rather than form a second phase. Coatings are not normally examined in this manner because of the complex nature of the sample.

Replication of the surface or a section of any tin-containing material is relatively easy. A carbon film is evaporated on to the surface in the usual manner and, since this will usually show little contour or phase contrast, shadowing is used. This involves evaporation of a heavy element (usually platinum, gold or gold—palladium) at a known angle and in a single known direction. The main problem arises in removing this fragile replica film from the sample without damage, since it often adheres to the surface quite strongly. Metallographic etching solutions may be used to dissolve the metal surface and hence separate the carbon replica, but no evolution of gas can be tolerated. Solutions which have been used successfully are given in Table 36.

## Table 36

Stripping solutions for removal of replicas

| Replicated specimen | Stripping solution (use at room temperature) |
|---|---|
| Most tin-rich alloys and surface of intermetallic compound layers (e.g. $FeSn_2$, $Cu_6Sn_5$, $Ni_3Sn_4$, etc.) | 20 ml/l HCl in methanol with 2 V applied potential |
| Surface of intermetallic compound layer in tinplate ($FeSn_2$) | 10–50 ml/l $Br_2$ in methanol |
| Bronze | 20 g/l $FeCl_3$, 50 ml/l HCl in a 2:1 mixture of Industrial Methylated Spirits and water |

Water-free methanol should always be used to wash replicas once they have been stripped from the sample.

In order to examine the surface of an intermetallic compound layer that has been formed by reaction of tin or a tin alloy (solid or liquid) with a basis metal (for example the coating/intermetallic layer interface), the tin coating must first be removed without damaging or altering the underlying intermetallic compound layer (see Fig. 86). This may be effected most easily by agitating the specimen in a solution of 35 g/l $o$-nitrophenol in 50 g/l NaOH at 60 to 70 °C for sufficient length of time. Thick tin–lead coatings (especially those of high lead content) usually require a longer treatment than pure tin coatings.

Thin film transmission microscopy of tin-rich alloys has been found to present problems since electrolytic thinning techniques appear to be difficult to control to produce the correct metal film thickness without complete disintegration of the sample.

### Scanning method

No problems exist for the examination of samples in the scanning electron microscope (SEM). Using a low incident angle for the electron beam with respect to the specimen, it is possible to illustrate clearly the geometric configuration of intermetallic compound crystals formed by solid state reaction (Fig. 113). It is interesting to compare the detail with that seen in the TEM picture (Fig. 86).

Due to the variation in absorption with atomic number, distribution of phases may often be observed without resorting to chemical etches, especially for example with tin–lead two-phase alloys. The technique allows non-destructive qualitative and quantitative analyses, using an X-ray spectrometer attachment to the SEM (Fig. 114).

Fig. 113 – Very low angle scanning electron microscope picture of Ni$_3$Sn$_4$ crystals formed by solid state diffusion in tin-plated nickel. The unrelated tin has first been removed by a sodium hydroxide/$o$-nitrophenol solution. (Magnification ×10 000.)

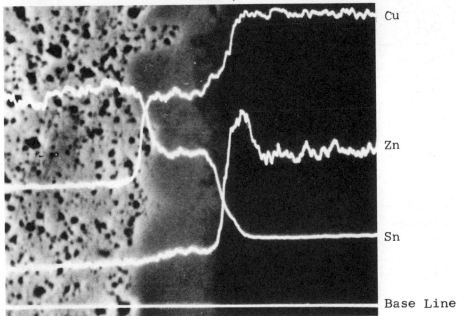

Fig. 114 – Line scan of X-ray intensities for the elements copper, tin, lead in a section of tin-lead alloy plated copper heated at 170 °C to cause solid state diffusion and intermetallic compound layer formation. (Magnification ×2000.)

**FURTHER READING**

1. *Metallographic Preparation of Tin and Tin Alloys,* International Tin Research Institute Publication No. 580.
2. *Metals Handbook,* 8th edn., Vol. 8 Metallography, structures and phase diagrams, American Society for Metals, Ohio, 1973.
3. Kehl, G. L., *The Principles of Metallographic Laboratory Practice,* 3rd edn., McGraw-Hill, New York, 1949.
4. Goodhew, P. J., *Specimen Preparation in Materials Science,* North-Holland Publishing Company, Amsterdam, 1973.
5. Phillips, V. A., *Modern Metallographic Techniques and their Applications,* Wiley, New York, 1971.

## 2. SI UNITS USED IN THE INDUSTRIAL METAL SERIES

| Quantity | SI unit | Recommended multiples | Other related units (or names) |
|---|---|---|---|
| Length | m (metre) | km<br>m<br>mm<br>nm | $Å = 10^{-10}$ m |
| Area | $m^2$ | $mm^2$ | hectare (ha) $= 10^4$ $m^2$ |
| Volume | $m^3$ | $mm^3$ | litre (l) $= 10^{-3}$ $m^3$ |
| Time | s (second) | s<br>ms<br>ns | minute (min)<br>hour (h)<br>day (d) |
| Mass | kg | g<br>mg<br>$\mu$g | tonne (t) $= 10^3$ kg<br>metric carat $= 2 \times 10^{-4}$ kg |
| Density | $\dfrac{kg}{m^3}$ | $\dfrac{g}{cm^3}$ | |
| Force | N (newton) | MN<br>kN | dyne $= 10^{-5}$ N |
| Impact strength | $\dfrac{J}{m^2}$ | $\dfrac{kJ}{m^2}$ | $\dfrac{J}{cm^2}$ |

| Quantity | SI unit | Recommended multiples | Other related units (or names) |
|---|---|---|---|
| Electric current | A (ampere) | mA | |
| Electric potential | V (volt) | MV kV mV | |
| Current density | $\dfrac{A}{m^2}$ | $\dfrac{A}{cm^2}$ $\dfrac{A}{mm^2}$ | |
| Resistance | Ω (ohm) | mΩ μΩ | |
| Resistivity | Ωm | μΩm | $\mu\Omega cm = 10^{-8}\ \Omega m$ |

## 3. COMMON CONVERSION FACTORS

| | |
|---|---|
| 1 yard | = 0.9144 m |
| 1 foot | = 0.3048 m |
| 1 inch | = 25.4 mm |
| 1 mil | = 0.001 in = 25.4 $\mu$m |
| 1 yd$^2$ | = 0.836 127 m$^2$ |
| 1 ft$^2$ | = 0.092 903 m$^2$ |
| 1 in$^2$ | = 6.4516 cm$^2$ |
| 1 yd$^3$ | = 0.764 555 m$^3$ |
| 1 ft$^3$ | = 28.3168 dm$^3$ |
| 1 in$^3$ | = 16.3871 cm$^3$ |
| 1 gal (Imp.) | = 4.5461 litre or = 4.54609 dm$^3$ |
| 1 in$^4$ | = 41.6231 cm$^4$ (Moment of Section) |
| 1 troy oz. | = 31.1035 g |
| 1 dwt (pennyweight) | = 1.555 17 g |
| 1 oz (av) | = 28.3495 g |
| 1 lb | = 0.453 592 kg |
| 1 cwt | = 50.802 kg |

1 ton                    = 1016.05 kg
1 tonne (metric)         = 1000 kg
1 lb/in$^3$              = 27.6799 g/cm$^3$
1 lbf                    = 4.448 22 N
1 tonf                   = 9964.02 N
1 tonf/in$^2$            = 15.4443 MN/m$^2$ or = 15.4443 N/mm$^2$
1 assay ton              = 32.6667 g

# Index

## W

Water repellency, 228
Wave soldering, 100, 102, 104
Welding, of tinplate, 192, 193
Wetting, of solders, 98
White tin, 13
Whitemetal, 110, 113
    alloy system, 47
    applications, 110, 115
    bearings manufacture, 119
Wipe tinning, 162, 170
    of bearing shells, 170
Wire, extruded, 50, 141
Wire drawing, tin and alloys, 146
Wood preservatives, 216, 222
Wood's metal, 58, 59
Work hardening, of tin, 17, 53
Work softening, of tin, 17, 54
Wrought bronze, 78
Wurtz process, 212

## X

X-ray data, 13, 14

## Y

Young's modulus, 17

## Z

Zinc, alloying with, 133
Zinc chloride (*see* Flux)
Zinc-tin
    alloy system, 46, 90
    electroplate, 157
Zircaloy, 232
Zirconium-tin alloys, 232